SpringerBriefs in Speech Technology

Studies in Speech Signal Processing, Natural Language Understanding, and Machine Learning

Series Editor

Amy Neustein, Fort Lee, NJ, USA

SpringerBriefs present concise summaries of cutting-edge research and practical applications across a wide spectrum of fields. Featuring compact volumes of 50 to 125 pages, the series covers a range of content from professional to academic. Typical topics might include:

- A timely report of state-of-the-art analytical techniques
- A bridge between new research results, as published in journal articles, and a contextual literature review
- A snapshot of a hot or emerging topic
- An in-depth case study or clinical example
- A presentation of core concepts that students must understand in order to make independent contributions

Briefs are characterized by fast, global electronic dissemination, standard publishing contracts, standardized manuscript preparation and formatting guidelines, and expedited production schedules.

The goal of the**SpringerBriefs in Speech Technology** series is to serve as an important reference guide for speech developers, system designers, speech engineers and other professionals in academia, government and the private sector. To accomplish this task, the series will showcase the latest findings in speech technology, ranging from a comparative analysis of contemporary methods of speech parameterization to recent advances in commercial deployment of spoken dialog systems.

** Indexing: books in this series are indexed in Scopus **

More information about this series at http://www.springer.com/series/10043

Krunal N. Patel

Robust and Secured Digital Audio Watermarking

Using a DWT-SVD-DSSS Hybrid Approach

 Springer

Krunal N. Patel
G.H Patel College of Engineering
and Technology, CVM University
Bakrol, India

ISSN 2191-737X   ISSN 2191-7388   (electronic)
SpringerBriefs in Speech Technology
ISBN 978-3-030-53910-8   ISBN 978-3-030-53911-5   (eBook)
https://doi.org/10.1007/978-3-030-53911-5

This Springer imprint is published by the registered company Springer Nature Switzerland AG
The registered company address is: Gewerbestrasse 11, 6330 Cham, Switzerland

Preface

Today, the use of digital data like image, audio, and video is tremendously increasing due to the advancement in technology and the Internet revolution. With these advancements, attaining one's ownership and copyrights for this digital data is the biggest challenge. Digital watermarking is one of the technique to attain one's ownership and copyrights securely. It is the technique in which the owner's copyright information can be embedded into the original media in the form of an image, audio, text, or video. There are two main factors we need to observe for digital audio watermarking to maintain the robustness as well as imperceptibility against piracy, malicious attacks, and various transformation operations. Though there are many challenges to achieve these results, there are two approaches presented for audio watermarking that are used to improve the robustness, and imperceptibility of the embedded information with security. To provide the security, DSSS Encryption and Decryption algorithm is used which is based on synchronized secret key concept along with DWT, DFT and SVD transformation methods. DWT (Discreet Wavelet Transformation) is used up to the fourth level to get the lowest frequency sub-band and after that DFT is applied to get the lowest frequency from the sub-band found by DWT in which the modifications are done, and then SVD (Singular Value Decomposition) is applied to it, so that the original audio file does not have any impact of the watermark bits to get the better robustness and imperceptibility.

Bakrol, India   Krunal N. Patel

Acknowledgments

First and foremost, praises and thanks to the God, the Almighty, for His showers of blessings throughout my research work to complete the research successfully.

I forward my sincere thanks to **Prof. Saurabh A. Shah** and **Prof. Dipti B. Shah** for their valuable help during the work of this book. Their suggestions were always there whenever I needed them. As a research supervisor, they spared their valuable time for the in-depth discussion on this research work. Also, I would forward my hearty thanks to the Department of Computer Science and Engineering and the department head **Dr. Chirag D. Patel** for providing us the computer laboratory access during our needs.

I am extremely grateful to my parents for their love, prayers, caring, and sacrifices for educating and preparing me for my future. Also, I express my thanks to my sister and brother for their support and valuable prayers.

Hearty thanks to **department administrators of C U SHAH University**, for their kind cooperation and providing us with necessary software whenever we asked for it.

Finally, I would like to thank all the people who were helpful to me, whether directly or indirectly for the completion of this research work.

Patel Krunalkumar N.

Contents

List of Figures

List of Tables

Chapter 1
Introduction

Abstract This book deals with the digital audio watermarking copyright assurance. The watermarking data that will be utilized incorporates text messages, copyright audio, written by hand marks, and cell phone numbers. The purpose behind this is they are more instructive contrasted with pseudo irregular groupings. The watermarking data will be encoded before embedding it in the audio. DSSS algorithm is used to increase the security of this system. The embedding process of the watermark information is performed in the frequency domain rather than the spatial domain. The combination of DWT, SVD, and DFT methods will be utilized for this reason. The watermarks ought to be imperceptible and make insignificant contortion to the host audio. The main objective of this research is to propose a new algorithm that can embed and extract the watermarking information and fulfill the above prerequisites. The execution of the newly proposed algorithm will be surveyed by testing them utilizing a group of various audio file types and against various attacks.

Keywords Digital watermarking · DWT · SVD · DSSS · DFT · Watermarking embedding · Watermarking extraction

1.1 Introduction

Today, the use of digital data like image, audio, and video is tremendously increased due to the advancement in technology and internet revolution. With this advancement to attain the one's ownership and copyrights for this digital data is the biggest challenge. Digital watermarking is one of the techniques to attain one's own and copyrights with securely. It is the technique in which the owner's copyright information can be embedded into the original media either in the form of an image, audio, text, or video. There are two main factors we need to observe for this digital audio watermarking to maintain the robustness as well as imperceptibility against the piracy, malicious attacks, and various transformation operations. Though there are many challenges to achieve these results, in this paper, our proposed audio watermarking technique is used to improve the robustness, imperceptibility of the embedded information with security. For security, in our proposed work, we are using synchronized secret key concept with DSSS encryption algorithm and two important powerful transformation methods used that are DWT (discrete wavelet transformation) up to four level to get lowest frequency sub-band and after that DFT is applied to get the lowest frequency from sub-band found by DWT in which the modifications are done and then SVD (singular value decomposition) is applied to it so that original audio file does not have any impact of watermark bits to get the better robustness and imperceptibility.

In the twenty-first century due to globalization and advancement in communication technology, the use of the internet is growing rapidly. Because of this immense use of online resource sharing and accessing, which may include copying and downloading are increasing day by day. Today, the sharing and copying of digital content like image, audio, and video files without losing the quality over the internet are becoming easier in the day-to-day life (Singh and Chadha 2013). In the tradition approach, people were using analog content to fulfill this requirement but that was a very costlier way to transmit the data. However, this issue has been resolved with the help of digital content. But afterward, with the ease of copying the content without losing the quality has led to the copyright owners to financial loss. By considering these contexts, protecting the intellectual belongings rights of such digital works against malicious user attacks and piracy has become a sweltering research topic that requires imperative solutions (Cox et al. 2008). Furthermore, recording medium and distribution networks for analog multimedia are very costlier. One of the most promising approaches that have seriously attracted many of the researchers in recent years is digital watermarking; in earlier days, it was unsafe to transmit the authentic digital information securely, but after that, some traditional methods like cryptography and steganography are found to accomplish this secure surrounding.

Steganography methods are not robust against various attacks or modification of data that might occur during transmission. Before these methods, people were using invisible ink concept with writing information, merging two images to create a new one to hide the information, drawing a standard painting with some small modifications, shearing the header information of the messenger in the form of a message, etc. (Johnson et al. 2000).

Cryptography is used to protect the important information which is going to be transferred; however, the drawback of this method is once the data are decrypted, it is hard to find where the modification of the information has been done. Today, with the rapid growth of sharing and accessing the information over the internet, many applications like audio recording and sharing of audio files, they are not considering copyright protection issue frequently due to its complexity. But after considering this copyright violation issues, digital watermarking approach becomes the most promising solution compared to cryptographic and steganography processes, against the tempering and alteration of intellectual belongings (Durvey and Satyarthi 2014).

Hence, there is a requirement of some efficient protection techniques that should be efficient, robust, and secure to prevent such malicious attacks and unintended activities. To fulfill this, some researchers have introduced "Watermarking." The applications of watermarking include ownership protection, proofing for authentication, air traffic monitoring, medical applications, military applications, etc. This technique of watermarking has extensive use in music industry (Shah et al. 2008).

A watermarking is a technique in which a watermark (that is a unique identifier) is embedded in image or audio or video, typically used to identify ownership of copyright holders. It is the process of embedding information into a signal (e.g., audio, video, or image) in a way that is difficult to remove or temper. If someone tries to copy the content, then the information is also carried in the copy. Watermarking has become very important to enable copyright protection and ownership verification and save the intellectual property from any unauthorized person/persons (Saini and Shrivastava 2014).

1.2 Motivation

Today, with the rapid growth of sharing and accessing the information over the internet many applications like audio recording and sharing of audio files, they are not considering copyright protection issue frequently due to its complexity. But after considering this copyright violation issue, digital watermarking approach becomes the most promising solution compared to cryptographic and steganography processes, against the tempering and alteration of intellectual belongings.

Different algorithms are proposed in the spatial domain and transform domains for digital watermarking. The techniques that are defined in spatial domain still not resistant enough to compression and other data processing and have a relatively low-bit capacity (Natgunanathan et al. 2012). For instance, a simple noise in the data may corrupt the embedded watermark data. Besides, the techniques defined for frequency domain can embed more bits for the watermark and are more robust to attack. The discrete cosine transform (DCT) and discrete wavelet transform (DWT) are used for watermarking in the frequency domain (Jain and Jain 2015).

Usually, any efficient watermarking technique must fulfill certain properties or objectives where the most important are robustness, imperceptibility, and complexity (Shore et al. 2013; Pahlavani and Pourmohammad 2013). Robustness refers to the case where the watermark is retained in the source content despite several stages of processing (Ai et al. 2013). Imperceptibility ensures that the quality of the host signal is not perceivably distorted (Dhar and Shimamura 2013). Lastly, complexity belongs to the amount of effort and time required for watermark embedding and extraction techniques (Seitz 2005). In considering the attacks on watermarks, the above feature of an algorithm becomes very important. According to that, the classification of a watermark algorithm is called robust if the watermark data embedded by that algorithm in an audio or any other digital data cannot be altered or removed easily. If anyone tries to do that then the data itself destroyed. To achieve this, comparative study has been performed on different watermark algorithms for their robustness. In fact, the robustness of the algorithms is dependent on the frequency at which the watermark data are added. Here in this research work, we have proposed the watermarking algorithm for audio watermarking along with encryption to make it secure.

1.3   The Content of the Book

This book deals with the digital audio watermarking copyright assurance. The watermarking data that will be utilized incorporates text messages, copyright audio, written by hand marks, and cell phone numbers. The purpose behind this is they are more instructive contrasted with pseudo irregular groupings. The watermarking data will be encoded before embedding it in the audio. DSSS algorithm is used to increase the security of this system. The embedding process of the watermark information is performed in the frequency domain rather than the spatial domain. The combination of DWT, SVD, and DFT methods will be utilized for this reason. The watermarks ought to be imperceptible and make insignificant contortion to the host audio. The main objective of this research is to propose a new algorithm that can embed and extract the watermarking information and fulfill the above prerequisites. The execution of the newly proposed algorithm will be surveyed by testing them utilizing a group of various audio file types and against various attacks.

The main objective of this research is to propose a new digital watermark algorithm that preserves the copyright property of the audio files. Here, the two most powerful techniques DWT and SVD are used with the combination of other techniques DFT and DSSS to enhance the security and also provide high robustness against the various malicious attacks.

This technique has unique features like:

- To implement a robust audio watermarking algorithm which will give the high robustness against various malicious attacks.
- It would also minimize the degradation ration of original content ineffective way.
- To provide the best SNR for provided attacks for all types of audio data.
- It also provides security using DSSS Encryption.
- It provides the reliability of extracted watermark by measuring BER.
- To make a secured algorithm which can be used in the defense system for transmitting the hidden message in audio signal.

The uniqueness of this research can be summarized as follows:

- Two popular techniques DWT and SVD are used with the combination of DFT.
- DSSS encryption algorithm is used to increase security.
- The proposed algorithm was implemented and tested against a variety of audio files.
- The important parameters like SNR (signal to noise ratio), BER (bit error rate) are evaluated to check the originality and quality of original audio data and watermark data.

After this introductory chapter, the following chapters of this book are organized in the following pattern:

Chapter 2 contains the introduction of digital watermarking, various types of digital watermarking, why watermarking is required, and various application areas where watermarking is useful. Various data hiding methods are also discussed. The basic principle of watermark embedding and watermark extraction is also explained. Watermarking classification depending on the various parameters and also the classification of watermarking are explained in this chapter.

Chapter 3 depicts the overall information about the various algorithms and techniques available for digital watermarking and also it covers the emerging issues in the existing algorithms. This chapter contains the working strategy of different spatial domain and frequency domain techniques like DWT (discrete wavelength transform), DCT (discrete cosine transform), SVD (singular vector decomposition), LSB (least significant bit). In this chapter, aside from giving more attention just to those papers, which are identified with this proposal work, the mind has been taken to cover increasingly up and coming ideas. After that, the deficiencies and the open doors for the exploration work are recognized and in view of those, inquire about issues that are produced by giving appropriate supports.

In Chap. 4, the problem statement is identified by exploring and studying various existing methods. In this research, the main objective is to provide the robust and secured watermarking algorithm, which can stand against the various malicious attacks by the attackers. And also it maintains the quality of the original audio data as well as the watermark data.

After discussing all these, our main research discussion starts from Chap. 5, in that digital audio watermarking algorithm is discussed. Initially, based on the survey done, the various methods available for digital audio watermarking are taken for consideration. From those methods, we found one approach that is relatively better than other approaches proposed. It is based on the combination of DWT and SVD methods. So, we have implemented that algorithm first and found that it is better but still there are some improvements needed in that to increase the robustness because we found that this algorithm is not fitted for some kind of attacks as well as some different types of watermark images. After that, we compare our results with existing methods and we found better results than other methods. But still, some improvements are needed in our work like in echo attack it is not giving the desired result. So, again we proposed another new approach that is the combination of DWT and SVD with DFT to increase the robustness and DSSS algorithm to provide the security to our algorithm. Our results show tremendous improvement against various malicious attacks and also improved the robustness.

Chapter 6 describes the results produced by the proposed algorithm and also performance analysis by comparing all the results with the existing algorithms has been discussed.

In Chap. 7, overall summary of the proposed work is described and also the scope of future work in this domain is discussed. At last in the bibliography section, all the papers which are referred for this work are listed and also the publication information for this research work is given.

References

Ai H, Liu Q, Jiang X (2013) Synchronization audio watermarking algorithm based on DCT and DWT. In: Proceedings of IEEE conference anthology, Shanghai, 1-8 January 2013. IEEE, New York, pp 1–4

Cox IJ, Miller ML, Bloom JA, Fridrich J, Kalker T (2008) Digital watermarking and steganography, 2nd edn. Elsevier, Amsterdam. ISBN: 978-0-12-372585-1

Dhar PK, Shimamura T (2013) Entropy-based audio watermarking using singular value decomposition and log-polar transformation. In: Proceedings of 2013 IEEE 56th international midwest symposium on circuits and systems (MWSCAS), Columbus, 4–7 August 2013. IEEE, New York, pp 122–127

Durvey M, Satyarthi D (2014) A review paper on digital watermarking. Int J Emerg Trend Technol Comput Sci 3(4):272

Jain R, Jain M (2015) Digital image watermarking using 3-level DWT and FFT via image compression. Int J Comput Appl 124(16):35–38

Johnson NF, Jajodia S, Duric Z (2000) Information hiding: steganography and watermarking attacks and countermeasures. Academic Publishers, Kluwer

Natgunanathan I, Xiang Y, Rong Y, Zhou W, Guo S (2012) Robust patchwork-based embedding and decoding scheme for digital audio watermarking. IEEE Trans Audio Speech Lang Process 20:2232–2239

Pahlavani F, Pourmohammad A (2013) A block set interpolation technique based additive-white-noise robust audio watermarking method. In: Proceedings of 2013 10th international ISC conference on information security and cryptology (ISCISC), Yazd, 29–30 August 2013. IEEE, New York, pp 1–5

Saini LK, Shrivastava V (2014) A survey of digital watermarking techniques and its applications. Int J Comput Sci Trend Technol 2(3):70

Seitz J (2005) Digital watermarking for digital media. Information Science, Hershey

Shah P, Choudhari P, Sivaraman S (2008) Adaptive wavelet packet based audio steganography using data history. In: Region 10 colloquium and the third ICIIS. IEEE, Kharagpur

Shore S, Ismail M, Zainal N, Shokri A (2013) Error probability in spread spectrum (SS) audio watermarking. In: Proceedings of 2013 IEEE international conference on space science and communication (IconSpace), Melaka, 1–3 July 2013. IEEE, New York, pp 169–173

Singh P, Chadha RS (2013) A survey of digital watermarking techniques, applications and attacks. Int J Eng Innovat Technol 2(9):165–175

Chapter 2
Path to Digital Watermarking

Abstract This chapter contains the introduction of digital watermarking, various types of digital watermarking, why watermarking is required, and various application areas where watermarking is useful. Various data hiding methods are also discussed. The basic principle of watermark embedding and watermark extraction is also explained. Watermarking classification depending on the various parameters and also the classification of watermarking is explained in this chapter.

Keywords Types of watermarking · Applications of watermarking · Digital watermarking · Watermarking classification · Parameters of watermarking algorithm

2.1 Introduction to Digital Watermarking

The powerfully expanding development of the web in the previous years has quickly expanded the accessibility of computerized information, for example, audio, pictures, content, and videos to the general population. Therefore, the issue of securing sight and audio data turns out to be especially critical. This issue can be solved using a well-known technique such as digital watermarking, which is the most widely recognized and perhaps most grounded strategy for ensuring advanced information.

Digital watermarking is a strategy to insert an intellectual copyrighted information in the electronic world. Digital watermarking is the way toward embedding implanted data into an original digital signal. That digital signal might be audio, image, or video. The data which are going to be embedded are known as watermark data that can be extracted or identified. A watermark might be a digital signal or an encrypted pattern which is embedded into a digital data such as audio. Since this signal or pattern is available in each unaltered duplicate of the original information, the watermark may likewise fill in as an advanced mark for the duplicates as a digital signature.

2.2   Why Digital Watermarking Is Required?

- For copyright protection for original content.
- To avoid the forgery.
- For security of the original content.
- To claim the ownership.
- It helps the owner to identify whether his/her multimedia data is stolen by others or not specifically from social media sites.

2.3   Fundamental Working Principle of Digital Watermarking Process

Here, in this section, we are going to discuss the basic working process of digital watermarking. Digital watermarking contains mainly two steps.

1. Watermark embedding process.
2. Watermark extraction process.

The first step is watermark embedding and the second step is watermark extraction. In watermark embedding process, the watermark data that may be any text or image are embedded in the original data. For increasing the security of the content, security key is also embedded in it for unauthorized access by third parties (Rashid 2016). By combining all this watermark embedding process is carried out. The basic watermark embedding process is depicted in Fig. 2.1.

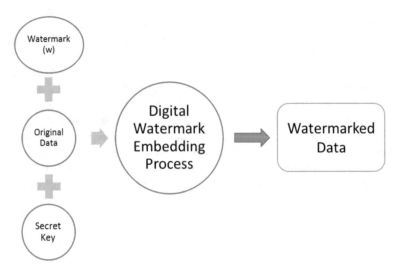

Fig. 2.1 Digital watermarking embedding process

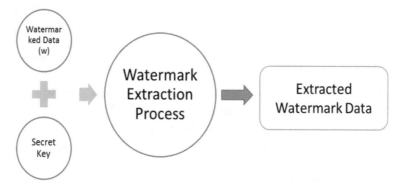

Fig. 2.2 Digital watermarking extracting process

Now, for the watermark extraction process, the above generated watermarked data are used as an input and the secret key is used for the authentication. After that successfully we can extract the watermark data from the digital watermark extraction process. The basic watermark process is depicted in Fig. 2.2.

2.4 Types of Digital Watermarking

2.4.1 According to Media Type

1. **Image watermarking:** Digital image watermarking is the technique in which watermark is going to be embedded in an image. There are various approaches to this method. It can be divided on the basis of the domain or transform or on the basis of wavelet (Usha and Kumar 2016). In spatial domain technique, work is directly been done on pixels and the frequency domain works on the basis of transformation coefficient. Using the property of the human visual system (HVS) and the image data is added with imperceptibility.
2. **Text watermarking:** In this type of watermarking technique, data are added to any text documents like PDF, DOC, and so on. Watermark can be added in such a way that it carries the secret message so that copyright can be easily recognized (Tiwari and Sharmila 2017).
3. **Audio watermarking:** As due to the popularity of music in youth there may be a chance of huge forgery in copyright music. To avoid this audio watermarking is a scorching issue for the researchers. In this technique, the watermark can be embedded in time or frequency domain in such a way that it will not affect the audibility of the original audio (Tiwari and Sharmila 2017).
4. **Video watermarking:** The illicit dissemination of movies is a typical and huge risk to the film business. With the coming of the rapid broadband Internet get to, a pilfered duplicate of a digital video would now be able to be effectively conveyed to a worldwide gathering of people. A conceivable method for restricting this sort

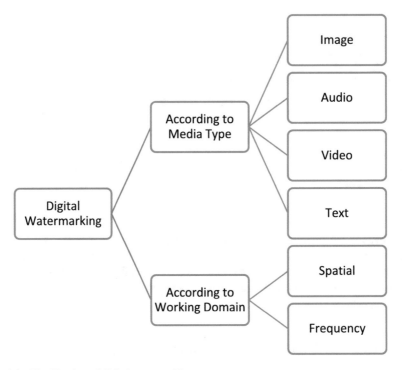

Fig. 2.3 Classification of digital watermarking

of advanced burglary is digital video watermarking whereby additional data, called a watermark, are inserted in the host video stream. This watermark can be extricated at the decoder and used to decide if the video content is watermarked (Fig. 2.3).

2.4.2 According to Working Domain

1. **Spatial domain:** This area centers around changing pixels of maybe a couple of haphazardly chosen subsets of pictures. In the spatial domain, we manage pictures as it seems to be. The estimation of the pixels of the picture change regarding scene/time spatial area strategy inserts the information by specifically adjusting the pixel estimations of the original picture. Some of the methods are the least significant bit (LSB), SSM modulation based, patchwork based.
2. **Frequency domain:** In the frequency area, the watermark is installing into recurrence coefficients of the host picture. Recurrence area watermarking gives more data concealing limit and high robustness against different geometrical assaults. Frequency domain watermarking is more robust than spatial domain watermarking because of the inserting of the watermark into the changed

Fig. 2.4 Basic watermarking process in spatial domain

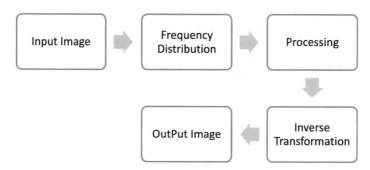

Fig. 2.5 Basic watermarking process in frequency domain

frequency coefficients of the transformed image (Mahmoud El-Gayyari et al. 2013) (Figs. 2.4 and 2.5).

This is also known as a transform domain technique. The most frequently used transform domain techniques are DCT (discrete cosine transform), DWT (discrete wavelet transform), and DFT (discrete Fourier transform).

2.5   Digital Audio Watermarking

With the expanding utilization of digital media, the assurance of licensed innovation rights issue has turned into a vital issue. Advanced watermarking is currently drawing consideration as another technique for shielding interactive media content from unapproved duplicating and to protect the data against unauthorized access (Komal et al. 2012). Audio watermarking is a strategy for covering the message into a sound to shield from unauthorized sources. On the off chance that any data are conveyed to somewhere else, at that point to shield this data from illicit replicating, it ought to be watermarked to give security to data. In militaries, remote correspondence which requires mystery correspondence among sender and beneficiary point (Singh and Singh 2015). Presently multiday on the grounds that the development of web developing is extremely fast, there might be probability to the unapproved conveyance of advanced information. When contrasted with picture and video watermarking, audio watermarking techniques are difficult because of human auditory system (HAS) is touchier than human visual system (HVS) (Elshazly et al. 2012). Because of low testing recurrence, a little measure of clamor in HAS framework can be recognized by ear. It is an exceptionally prevalent research territory in advanced media information stowing away (Ren and Li 2011).

2.6   Applications of Digital Watermarking

Watermarking techniques can be extremely helpful in a few territories of intrigue. The prerequisites that a watermarking technique needs to conform to rely on the particular kind of use. According to that, there are some common uses of this are as follows according to Arnold (2000).

2.6.1   Copyright Protection

The proprietor identification can be composed of the spreads or said someplace on the protest. For instance, a music company turns out with its CD/DVD by the recognizable proof characteristic of a sound organization on the CD case or the sign of the papermaker on the top corner of the paper (Arnold 2000). These sorts of watermarks can be essentially isolated by picking the picture or by tearing the part that has recognizable proof. Computerized watermarking encourages to crush this issue by embedding the watermark as bits and shaping a fundamental piece of the substance.

2.6.2   Authentication

Digital information can be watermarked to show that the computerized content cannot be unlawfully repeated. Devices equipped for replication would then be able to identify such watermarks and anticipate unapproved replication of the substance.

2.6.3   Tamper-Proofing

In this application, digital watermark data are inserted in the original signal and can be utilized to check if the original signal is altered. This circumstance is essential since it is important to think about the altering caused to the audio signal. The altering is at some point a reason for manufacturing of the watermark which must be stayed away from. If the watermark extracted is used for authentication of the original content (Shete and Kolkure 2016; Kundur and Hatzinakos 1999).

2.6.4   Broadcast Monitoring

It is imperative for creating organizations to counteract illicit rebroadcasting exercises. For this situation, digital watermarks can be utilized to naturally screen broadcasting streams at satellite hubs everywhere throughout the world and distinguish any unlawful communication material. Also, numerous associations are keen on secure techniques for getting the greater part of the broadcast appointment they buy from supporters (Hofbauer and Hering 2007). Procedures for registering can be ordered with two sorts: inactive checking that endeavors to specifically perceive the substance being communicated and dynamic observing which depends on related data that are communicated alongside the substance.

2.6.5   Fingerprinting

A fingerprinting strategy, for the most part, used to delineate the wellspring of unlawful duplicate. Each duplicate accessible can be watermarked with a solitary piece grouping. For the proprietor, embedding a novel serial number-like watermark is a fine method to detect clients who break their permit assertion by duplicating the ensured information and providing it to an outsider (Hussein and Belal 2012).

2.6.6   Media Forensics

In forensic watermark, applications upgrade a substance proprietor's capacity to identify and react to abuse of its advantages. Measurable watermarking is utilized to assemble prove for criminal procedures as well as to uphold authoritative use assent ions between a substance proprietor and the general population or organizations with which it shares its substance (Kundur and Hatzinakos 1999).

2.6.7   Medical Application

In these watermarking strategies, the names of the patients can be imprinted on the X-ray reports and MRI examines. The medical reports assume an extremely critical part in the treatment offered to the patient. In the event that there is a misunderstanding in the reports of two patients this could prompt a disaster. Consequently, embedding the date and patient's name in X-ray or MRI pictures could expand the secrecy of restorative data and also the security.

2.6.8   Picture Authentication

In a picture verification application, the expectation is to identify adjustments to the information. The qualities of the picture, for example, its edges, are installed and contrasted and the present pictures for contrasts. An answer to this issue could be acquired from cryptography, where the advanced mark has been examined as a message verification strategy. One case of computerized signature innovation being utilized for picture validation is the reliable advanced camera.

2.6.9   Airline Traffic Monitoring

Watermarking is utilized as a part of air movement observing. The pilot speaks with a ground-observing framework through voice at a specific frequency. In any case, it can be effectively caught and assaulted and is one of the reasons for miss correspondence. To maintain a strategic distance from such issues, the flight number is installed into the voice correspondence between the ground administrator and the flight pilot. As the flight numbers are one of a kind, the following flights will turn out to be more secure and simple.

2.6.10   Communication of Ownership

Computerized content keeps on multiplying as the present customers look for data and diversion on their PCs, cell phones, and other advanced gadgets. In our digital culture, advanced has turned into an essential method for correspondence and articulation. The mix of access and new apparatuses empowers advanced substance to movement speedier and more remote than any time in recent memory as it is transferred, scattered, seen, downloaded, altered, and repurposed at an amazing velocity. Regardless of whether you are a worldwide media partnership or an independent picture taker, the capacity to convey your copyright possession and use rights is basic (Podilchuk and Delp 2001).

2.7   Properties of Efficient Digital Watermarking

The efficiency of any digital watermarking techniques can be described by various properties. The relative significance of every property, by and large, relies upon the requests of the application. Based on that most significant properties are as follows:

2.7.1 Robustness

The robustness of a watermarking algorithm is characterized as its capacity to recognize/separate the watermark after normal signal handling controls. The arrangement of signal preparing adjustments to which a watermarking algorithm should be hearty against is totally application subordinate. For instance, in radio communication, inserted watermark requires just to survive bends caused by the transmission procedure, including dynamic pressure and low pass separating, in light of the fact that the watermark identification is done specifically from the communicated signal. Then again, in a few calculations, strength is totally unfortunate and those calculations are marked delicate audio watermarking algorithm (Patil and Chitode 2013; Elshazly et al. 2012). There are two noteworthy issues when attempting to assurance robustness, the watermark must be as yet exhibit in the media after the change or it must be as yet feasible for the watermark finder to identify it.

2.7.2 Perceptibility

One of the imperative highlights of the watermarking system is that the watermarked data ought not to lose the nature of the original signal. The signal to noise ratio (SNR) of the watermarked data to the original signal ought to be kept up more noteworthy than 20 dB. Furthermore, the strategy should make the changed signal not detectable by human ear. In all practically every application, the watermark-embedding algorithm needs to embed watermark information without changing the perceptual nature of the host audio signal. The constancy of a watermarking algorithm is typically characterized as a perceptual similitude between the first and watermarked audio succession. Be that as it may, the nature of the watermarked audio may get spoiled, either deliberately by a foe or inadvertently amid the transmission procedure, before a man sees it (Malshe (Gondhalekar et al. 2012). In such a case, it is more sensible to rethink the loyalty of a watermarking algorithm as a perceptual comparability between the watermarked audio and the first host audio at the time when they are exhibited to a shopper. Closeness factor is utilized to quantify perceptual straightforwardness of unique watermark and recovered watermark (Katzenbeisser and Petitcolas 2000).

2.7.3 Reliability

Reliability covers the highlights like the robustness of the audio signal against the wicked assaults and signal processing methods. The watermark ought to be made in a way that they give high strength against assaults (Xuehu et al. 2010). Likewise, the watermark identification rate ought to be high under any kind of assault in the circumstances of demonstrating proprietorship.

2.7.4   Security

Watermark algorithm must be secure to the point that an attacker must not have the capacity to distinguish the nearness of embedded information, not to mention evacuate the inserted information. The security of the watermark process is deciphered similarly as the security of encryption procedures and it cannot be broken except if the approved client approaches a secret key that controls watermark embedding. An unapproved client ought to be not able to concentrate the information in a sensible measure of time regardless of whether he realizes that the host signal contains a watermark and knows about the correct watermark-embedding algorithm (Xuehu et al. 2010). Security prerequisites shift with the application and the most stringent are in cover correspondences applications, and, now and again, information is encoded before embedding into the original audio signal.

2.7.5   Computational Cost and Complexity

The execution of the watermarking algorithm is a dull task and it relies upon the business application included. The primary issue from the specialized perspective is the computational complexity nature of inserting and location calculations and the quantity of embedded and finders utilized as a part of the framework. For instance, in a broadcast communication system, implanting and identification must be done progressively, while in copyright insurance applications; time is certifiably not a significant factor for a useful usage by Muharemagic and Furht (2001). One of the monetary issues is the plan of embedder and identifiers, which can be executed as equipment or programming modules, is the distinction in handling intensity of various gadgets (mobile phones, wireless phones, radio, etc.).

2.7.6   Speed

Watermark embedding speed is one of the criteria for proficient watermarking strategy. The speed of inserting the watermark is critical progressively applications where the embedding is done on nonstop signals, for example, the discourse of an authority or discussion between plane pilot and ground control staff or in military communication. A portion of the conceivable applications where speed is a limitation is sound spilling and aircraft activity checking. Both embedding and extraction process should be made as quick as conceivable with more noteworthy proficiency.

2.7.7   Capacity

The productive watermarking procedure ought to have the capacity to convey more data yet ought not to corrupt the nature of the original audio signal. The capacity of the watermarking framework is characterized as the most extreme measure of data that can be implanted in the cover data. The quantity of watermark bits in a message in information payload and the most extreme redundancy of information payload inside a picture is the watermark limit (Chaw-Seng 2002). It is likewise vital to know whether the watermark is totally dispersed over the host signal since it is conceivable that close to the extraction procedure a piece of the signal is just accessible. Subsequently, capacity is likewise an essential worry in the ongoing circumstances. A watermark may have a high information limit yet low information payload.

2.8   Performance Evaluation Factors

To check robustness and perceptual transparency we can use signal-to-noise ratio (SNR), robustness (ρ), and bit error rate (BER).

2.8.1   Signal-to-Noise Ratio

SNR is a statistical difference metric which is used to measure the similitude between the undistorted original audio signal and the distorted watermarked audio signal. The SNR computation is done according to Eq. (2.1) (Darabkh 2014), where A corresponds to the original signal and A' corresponds to the watermarked signal.

Although SNR is a simple metric to measure the noise introduced by the embedded watermark and can give a general idea of imperceptibility, it does not take into account the specific characteristics of the human auditory system (Beerends and Stemerdink 1992).

$$\text{SNR(dB)} = 10 \log 10 \frac{\sum_n A_n^2}{\sum_n (A_n - A\prime_n)^2} \qquad (2.1)$$

2.8.2   Robustness

Watermarked audio digital signals may undergo common signal processing operations such as linear filtering, lossy compression, among many others. Although these operations may not affect the perceived quality of the host signal, they may corrupt the watermark image embedded within the signal. To evaluate the robustness of the proposed algorithm (Darabkh 2014), we implemented a set of attacks that commonly affect audio signals. Most of these attacks have been defined (Al-Haj et al. 2011):

$$\rho(w, \hat{w}) = \frac{\sum_{1}^{n} wi\hat{w}}{\sqrt{\sum_{i=1}^{n} wi^2 \sum_{i=1}^{n} \widehat{wi^2}}} \qquad (2.2)$$

2.8.3 Bit Error Rate

Robustness is measured using the bit error rate (BER) metric since the watermark used in the simulation is a binary image. BER is defined as the ratio of incorrectly extracted bits to the total amount of embedded bits, as expressed in the following equation:

$$\text{BER} = \frac{100}{l} \sum_{n=0}^{i-1} \left\{ \begin{array}{l} 1, W'_n = W_n \\ 0, W'_n \neq W_n \end{array} \right\} \qquad (2.3)$$

where l is the watermark length, W_n is the nth bit of the embedded watermark, and W'_n is the nth bit of the extracted watermark. Reliability is measured using BER (bit error rate).

2.9 Different Types of Attacks

The vital necessities of a proficient watermarking algorithm are the robustness and inaudibility. There is an exchange off between these two necessities; in any case, by testing the watermark algorithm with the audio signal processing assaults that hole can be made insignificant. Each application has its particular prerequisites and gives an alternative to pick high strength repaying with the nature of the signal and the other way around. With no changes and assaults, each watermarking method performs productively. Probably the most widely recognized kinds of procedures an audio signal experiences when transmitted through a medium are as per the following:

2.9.1   Noise

It is normal practice to see the nearness of noise in a signal when transmitted. Henceforth, the watermarking algorithm should make the system strong against the commotion assaults. It is prescribed to check the calculation for this kind of noise by including the host signal by an additive white Gaussian noise (AWGN) to check its robustness (Steinebach et al. 2001).

2.9.2   Cropping Attack

Cropping is another technique that might be utilized by an enemy to change the perspective proportion without extending the signal or filling the clear spaces. An editing assault may influence the watermark with little change to the cover audio. That is relying upon the watermark strategy and its power. The cropping might be executed on some portion of the original signal. The editing assault expels a few segments of the watermarked picture, decreasing the extent of the host signal, which may make a portion of the shrouded data be lost. It is significant that editing additionally causes a synchronization issue for information concealing plans (Steinebach et al. 2001).

2.9.3   Dynamics

The amplitude change and the attenuation change are the progressions of the assaults. Capacity, expansion, and compression are a type of more convoluted applications which are the non-direct adjustments. A portion of these sorts of assaults is re-quantization (Tewari et al. 2011).

2.9.4   Amplify

The watermarked audio is intensified at various enhancement rates. Through this assault, the watermarking plans that install the watermark in the adequacy of the individual examples end up powerless as the amplitude is adjusted.

2.9.5 Filtering

Filtering is common practice, which is used to amplify or attenuate some part of the signal. The basic low pass and high pass filters can be used to achieve these types of attacks (Tewari et al. 2011).

2.9.6 Adding the Echo

Echoes with various deferrals are added to the watermarked audios. This assault fundamentally is the counter for echo addition conspires that includes the watermark bit as echoes with various deferrals.

2.9.7 Ambiance

In a few circumstances, the audio signal gets deferred or there are circumstances wherein an individual's record motion from a source and claim that the track is theirs. Those circumstances can be mimicked in a room, which is of incredible significance to check the execution of an audio signal (Tewari et al. 2011).

2.9.8 Removal Attack

Removal assaults expect to expel the watermark information from the watermarked audio signal. Such assaults misuse the way that the watermark is typically an added additive noise signal introduced in the host signal (Tewari et al. 2011).

References

Al-Haj A, Mohammad A, Bata L (2011) DWT-based audio watermarking. Int Arab J Informat Technol 8:326–333

Arnold M (2000) Audio watermarking: features, applications, and algorithms. In: IEEE international conference on multimedia and expo (II). Citeseer, Princeton

Beerends JG, Stemerdink JA (1992) A perceptual audio quality measurement based on a psychoacoustic sound representation. J Audio Eng Soc 40:963–972

Chaw-Seng W (2002) Digital image watermarking methods for copyright protection and authentication

Darabkh KA (2014) Imperceptible and robust DWT-SVD-based digital audio watermarking algorithm. J Softw Eng Appl 7:859–871. https://doi.org/10.4236/jsea.2014.710077

Elshazly AR, Fouad MM, Nasr ME (2012) Secure and robust high quality DWT domain audio watermarking algorithm with binary image. In: Computer engineering & systems (ICCES), 2012 seventh international conference on. IEEE, New York

Hofbauer K, Hering H (2007) Noise robust speech watermarking with bit synchronisation for the aeronautical radio, LNCS 4567. Springer, Berlin, Heidelberg, pp 252–266

Hussein E, Belal MA (2012) Digital watermarking techniques, applications and attacks applied to digital media: a survey. Int J Eng Res Technol 1(7):1

Katzenbeisser S, Petitcolas FAP (2000) Information hiding techniques for steganography and digital watermarking. Artech House, Inc., Boston, London

Komal V, Goenka, Pallavi K Patil (2012) Overview of Audio Watermarking Techniques, International Journal of Emerging Technology and Advanced Engineering, ISSN 2250-2459, Volume 2, Issue 2, February 2012

Kundur D, Hatzinakos D (1999) Digital watermarking for telltale tamper proofing and authentication. In: Proceeding of the IEEE. IEEE, New York, pp 1167–1180

Muharemagic E, Furht B (2001) A survey of watermarking techniques and applications

Patil M, Chitode JS (2013) Improved technique for audio watermarking based on discrete wavelet transform. Int J Eng Adv Technol 2(5):511

Podilchuk CI, Delp EJ (2001) Digital watermarking: algorithms and applications. In: IEEE signal processing magazine, July 2001. IEEE, New York

Rashid A (2016) Digital watermarking applications and techniques: a brief review. Int J Comput Appl Technol Res 5(3):147–150

Ren K, Li H (2011) Large capacity digital audio watermarking algorithm based on DWT and DCT. In: International conference on mechatronic science, electric engineering and computer August 19–22, 2011. Jilin, China, pp 1765–1768

Shete MA, Kolkure VS (2016) A review on digital watermarking, its features, need and various techniques. Int J Res Appl Sci Eng Technol 4

Singh BK, Singh AK (2015) Data hiding in audio using matlab software. Int J Adv Eng Sci Technol 2(3):333–340

Steinebach M, Petitcolas F, Raynal F, Dittmann J, Fontaine C, Seibel S et al (2001) Stirmark benchmark: Audio watermarking attacks. In: Proceedings of the international conference on information technology: coding and computing. Nevada, Las Vegas, pp 49–54

Tewari TK, Saxena V, Gupta JP (2011) Audio watermarking: current state of art and future objectives. Int J Dig Cont Technol Appl 5(7):306–314

Tiwari N, Sharmila (2017) Digital watermarking applications, parameter measures and techniques. Int J Comput Sci Netw Secur 17(3):184

Usha C, Kumar SR (2016) Digital image watermarking techniques and applications: a survey. Int J Adv Res Comput Sci Softw Eng 6:3

Xuehua J (2010) Digital watermarking and its application in image copyright protection. In: 2010 international conference on intelligent computation technology and automation. IEEE, New York

Chapter 3
Existing Methods of Digital Watermarking

Abstract This chapter depicts the overall information about the various algorithms and techniques available for digital watermarking and it also covers the emerging issues in the existing algorithms. This chapter contains the working strategy of different spatial domain and frequency domain techniques like DWT (discrete wavelength transform), DCT (discrete cosine transform), SVD (singular vector decomposition), and LSB (least significant bit). In this chapter, aside from giving more attention just to those papers, which are identified with this proposal work, the mind has been taken to cover increasingly up and coming ideas. After that, the deficiencies and the open doors for the exploration work are recognized and in view of those, inquiries about issues are produced by giving appropriate supports.

Keywords DWT · DCT · SVD · LSB

3.1 Existing Methods of Audio Watermarking

Existing methods of audio watermarking techniques are divided into two parts:

1. Spatial domain
2. Transform domain

3.2 Existing Methods of Audio Watermarking in Spatial Domain

Spatial domain advanced watermarking algorithms directly used the pixel value to be altered or manipulated and for that, it straightforwardly stacks the basic information into the original image (Xuehua 2010). Spatial domain watermarking can likewise be connected utilizing shading bands. Along these lines, the watermark shows up in just a single of the shading groups. This renders the watermark with the end goal that it is hard to recognize under general survey. The spatial area is

© The Author(s), under exclusive license to Springer Nature Switzerland AG 2021
K. N. Patel, *Robust and Secured Digital Audio Watermarking*, SpringerBriefs in
Speech Technology, https://doi.org/10.1007/978-3-030-53911-5_3

controlling or changing an image speaking to a protest in space to upgrade the image for a given application. Methods depend on direct control of pixels in an image. There are some commonly used algorithms in the spatial domain as follows:

3.2.1   Least Significant Bit (LSB) Technique

One of the most in good time procedures examined in the data stowing away of digital audio (and additionally other media writes) is LSB coding. In this approach, the least significant bit of each byte is replaced or manipulated which carries the information (Bender et al. 1996). In this procedure, LSB of a paired grouping of each sample of the digital audio document is supplanted with a paired secret message. This particular method is most commonly used for image watermarking because every pixel is represented to as a number consequently it will be anything but difficult to supplant the bits. Here, robustness is measured as the number of bits is altered in the original signal (Arnold et al. 2003). An indistinguishable method from utilized as a part of image watermarking can't without much of a stretch be exchanged to the audio watermarking domain in light of the fact that the human visual framework isn't as delicate to minor varieties as the human auditory system (Gopalan 2003). The included information would probably be distinguishable to the end client or an unintended interceptor. However, there is cause for worry in utilizing a procedure that may function admirably in one area, and exchanging it to a domain where it would seem, by all accounts, to be essentially not so much success but rather more liable to help unapproved discovery. The 2-bit LSB encoding scheme is depicted in Fig. 3.1.

3.2.2   Patchwork Technique

Patchwork technique is the statistical model for information hiding based on pseudorandom. The watermarked information is divided into two subsets randomly. In this method, patchwork subtly embeds a watermark with a specific measurement

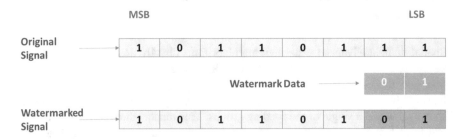

Fig. 3.1 LSB-embedding scheme

utilizing a Gaussian dissemination. A pseudo arbitrarily determination of two patches is done where the first is A and the second is B. Patch A picture information is lit up whereas that of patch B is obscured (for motivations behind this outline this is amplified). Patchwork being statistical strategies utilizes excess example encoding to embed message inside an image (Potdar et al. 2005). Here this particular method is very effective for image watermarking. It can survive against most of the alterations like rotation, cropping, scaling, and so on. Mix with the host signal depends on straightforward activities, in the pixel domain. The watermark can be distinguished by associating the normal example with the original signal. However, once the algorithm was found, it would be simple for a middle of the road gathering to adjust the watermark. One of the real restrictions in spatial domain technique is the limit of an image to hold the watermark (Mistry 2010).

3.2.3   Echo Hiding

Echo hiding method is covering up information and inserts into a unique audio signal by presenting an echo in the time domain with the end goal for straightforwardness. Here, binary messages are implanted by adding echo into the unique signal with one of two delays, either a D0 sample delay or a D1 sample delay. Extraction of the inserted message includes the location of delay D. Autocepstrum or cepstrum distinguishes the delay D. Cepstrum examination copies the cepstrum motivations each D samples. This method up is typically intangible and once in a while makes the audio ironic (Rathee and Kumar 2014). Synchronization strategies much of the time receive this strategy for coarse synchronization. An inconvenience of echo hiding method stowing away is its high many-sided quality due to cepstrum or autocepstrum calculation amid discovery. On the other hand, anyone can identify an echo with no earlier learning. As it were, it gives the insight for the vindictive assault. This is another detriment of echo hiding method by Bassia and Nikolaidis (2001) (Fig. 3.2).

Fig. 3.2 Echo-hiding watermarking scheme (Katzenbeisser and Petitcolas 2000)

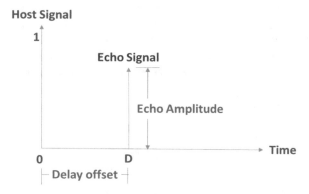

3.3   Existing Methods of Audio Watermarking
in Transform Domain

Transform domain watermarking scheme and introduces to the installing of the watermark information after the cover audio has been changed somehow. Lately, there has been an expanding number of research publications of watermarking techniques in view of control in a changing area or some likeness thereof, including yet absolutely not restricted to the discrete Fourier transform (DFT), discrete cosine transform (DCT), and discrete wavelet transform (DWT).

Here, in the below section, there is a brief introduction of the above techniques is provided.

3.3.1   Discrete Fourier Transform (DFT)

The discrete Fourier transform implies whereby the relative qualities of the different parts (frequencies) innate in a given signal can be computed (Kim et al. 2003). The "discrete Fourier transform" (DFT) can be actualized utilizing a "fast Fourier transform" (FFT), a quick execution of the DFT, on a sampled signal. The transformation plays out this count on discrete portions (frame) of the signal, each one in turn, regarding them as on the off chance that they were endlessly intermittent with a specific end goal to fulfill the prerequisites of Fourier ideas (FFT window functions: limits on FFT analysis 2009). The consequence of the DFT communicates the signal as far as complex exponential, catching data about the magnitude and phase of its intrinsic parts over the entire frame. The yield of the transformation on a given audio signal is the frequency interpretation of the signal, however, just for the casing under thought as though it was totally free of whatever is left of the signal. Where the frequency changes persistently, or at times however for a brief timeframe, the Fourier transformation is insufficient (Tan et al. 2009). At last, since it gives data just sufficiently about the parts to have the capacity to imitate the signal there may well be parts in the signal for which no data are caught. These contemplations confine the utilization of the FFT for signals, for example, music and recommend that expanded value is accomplished by utilizing littler and littler time portions (windows) subsequently expanding determination of the investigation yield, yet, in addition expanding computational intricacy. By the by, the change of a signal by DFT is an initial phase in numerous watermarking plans (Singh et al. 2009).

3.3.2 Discrete Cosine Transform (DCT)

The discrete cosine transform is the spectral transformation, which has the properties of discrete Fourier transformation (Watson 1994). The discrete cosine transform is like the DFT. Like the DFT, it states to or decays a signal with the goal that its parts may be depicted. Not at all like the DFT, the DCT utilizes as it were cosine waves. This is an effective calculation in light of the fact that the DCT is equivalent to a DFT or roughly double the length thus it is a critical advancement as far as computational unpredictability in signal processing undertakings. As there are numerous kinds of DCT with various info and parameter esteems, the DCT is adaptable in its applications.

The DCT has a large number of indistinguishable favorable circumstances from the DFT yet, in addition, huge numbers of similar impediments identified with nonstationary signals. In any case, its many-sided quality is significantly less than that of the DFT so considerably littler windows can be examined at the same computational cost (Yan et al. 2009). Note additionally that the DCT change is more valuable and more predominant in the change of image and video signals than the audio signal. The low-frequency DCT coefficients convey a considerable measure of energy and change in the lower frequency DCT coefficients prompt more prominent mutilation. And if the higher frequency coefficients are modified, then it will not create much more distortion but again there is a concern of robustness (Ali Khayam 2003).

3.3.3 Discrete Wavelet Transform (DWT)

Transformation of the audio signal from the time domain to frequency domain empowers watermarking algorithm to insert the watermark into a perceptually huge piece of a signal, which gives high strength against general signal processing activities. Wavelet transform gives them both time and frequency depiction of the signal. Discrete wavelet transform gives the adequate data to both analysis and synthesis of signal and is less demanding to execute (Wang and Zhao 2006). The term wavelet is an oscillatory vanishing wave with time-constrained broadens appeared in Fig. 3.3. By and large, wavelets are for the most part utilized as a part of signal processing applications. The analysis of nonstationary signals likewise should be possible with the assistance of wavelets (Xiaojuan et al. 2007).

The general one-level transformation of discrete wavelet transform decomposition steps is depicted in Fig. 3.4 to isolate high pass and low pass parts. Along these lines, the process includes time domain signal s[n] through a high pass channel also, down inspecting the signal acquired gives detailed coefficients (D). At the point when signal s[n] is gone through low-pass channels and down examining task gives approximate coefficients (A) (Polikar 2010).

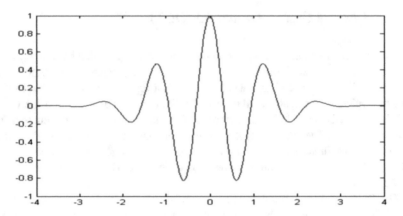

Fig. 3.3 Wavelet

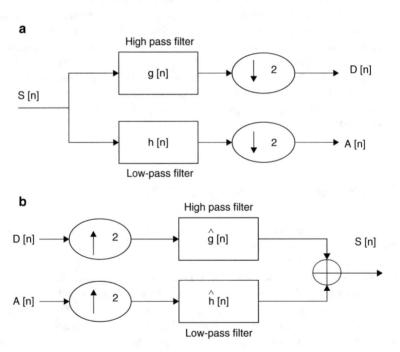

Fig. 3.4 One-level DWT signal decomposition and reconstruction (**a**) one-level DWT signal decomposition and (**b**) one-level DWT signal reconstruction (Gopalan 2003)

For most of the time signal decomposition segments chosen are low frequency coefficients. The "detail" coefficients are less vital for inserting watermarks since they contain less critical parts and are in this way ineffective against assaults. Contingent on the kind of use and length of the audio signal, the low-frequency part may be further disintegrated into two sections of high and low frequencies. The reproduction procedure is backward of the disintegration process. The approximate

Fig. 3.5 DWT coefficients
detail

LL [Approximate band]	HL [Horizontal band]
LH **[Vertical Band]**	HH [Diagonal band]

coefficients are up-sampled and went through a low-pass channel; comparatively, detailed coefficients are up-tested and gone through the high-pass channel. The got tests from these channels are convoluted to get the recreated signal of s[n] (Shahadi et al. 2014).

There is an exchange off between the robustness and the capacity in figuring out which level of DWT coefficients ought to be controlled. Controlling a lower level of coefficients gives higher capacity yet a lower robustness. Controlling a higher level of coefficients is helpful for accomplishing a higher robustness, however, a lower capacity. Also, a more elevated amount DWT includes a more computational cost. There are extraordinary sorts of DWT's accessible relying upon the kind of chose premise work (Chitode and Patil 2013).

Depending on that the Haar wavelet transform (HWT) is the least difficult of all wavelet transform functions. It is the fundamental orthogonal wavelet filter. In this, the low-frequency wavelet coefficient is produced by averaging furthermore, high-frequency coefficients are created by taking half of the distinction. The four groups acquired are an approximate band (LL), vertical band (LH), the horizontal band (HL), and inclining point of interest (detail) band (HH) (Fig. 3.5).

Here, the lowest frequency component is consisting of the approximation band which has the significant portion and the rest are three high-frequency components that are the detail portions.

The critical property of Haar Wavelet is that any genuine capacity can be approximated. The channel configuration incorporates two parts in this manner usage is simple. The vanishing minutes for Haar wavelet are 1 and are the essential wavelet. Haar wavelet is broadly utilized as a part of compression applications because of its basic wavelet and scaling capacities. DWT-based frequency domain watermarking technique, for the most part, utilized when we need to exchange more private issue through the web to anybody, in a military application, government application, communicate checking, that is, amusement and notices, and banks applications than DCT (Singh 2011).

3.4   SVD (Singular Value Decomposition)

Like DFT, DWT, and DCT, SVD transform method additionally discover its place in audio watermarking after it was effectively connected on image watermarking. This strategy gives a rich route for extracting logarithmic highlights from a 2-D matrix. The primary properties of the matrix of the SVs can be abused in watermarking. At the point, when a little bother happens to the first information matrix, no extensive varieties happen in the matrix of the SVs, which make this strategy strong against assaults (Özer and Sankur 2005; Zhang and Li 2005).

The SVD is worked on the metrics, so before applying SVD to audio the audio track is changed into a 2D matrix. The SVD of $M \times N$ matrix, X is characterized by the following equation (Sun et al. 2008).

$$SVD(X) = USV^T \qquad (3.1)$$

Where:

- A is an $M \times N$ matrix
- U is an $M \times R$ orthogonal matrix
- S is an $R \times R$ diagonal matrix
- V is an $R \times N$ orthogonal matrix

$$
\underset{m \times n}{\overset{X}{\begin{pmatrix} x_{11} & x_{12} & \cdots & x_{1n} \\ x_{21} & x_{22} & \cdots & \\ \vdots & \vdots & \ddots & \\ x_{m1} & & & x_{mn} \end{pmatrix}}} = \underset{m \times r}{\overset{U}{\begin{pmatrix} u_{11} & \cdots & u_{1r} \\ \vdots & \ddots & \\ \vdots & & u_{mr} \end{pmatrix}}} \underset{r \times r}{\overset{S}{\begin{pmatrix} s_{11} & 0 & \cdots \\ 0 & \ddots & \\ \vdots & & s_{rr} \end{pmatrix}}} \underset{r \times n}{\overset{V^T}{\begin{pmatrix} v_{11} & \cdots & v_{1n} \\ \vdots & \ddots & \\ v_{r1} & & v_{rn} \end{pmatrix}}} \qquad (3.2)
$$

The SVD operation divides a matrix into three orthogonal matrixes U, V, and S. The matrix S is an $(N \times N)$ diagonal matrix in which all the entries are zero except the diagonal. The diagonal elements thus produced are always in a descending order. The nonzero entries corresponding to the S matrix are called singular values (Bhat et al. 2011).

SVD is utilized for the most part for image watermarking and not many have connected it on audio watermarking. The SVD-based watermarking procedures are sorted into two groups—One utilizing the host signal for watermark extraction/ discovery called as the non-blind procedures, furthermore, the second classification that is blind which doesn't require the host signal.

In a portion of the watermarking plans utilizing the SVD, the unitary and also the particular singular matrix is to be embedded till the procedure of extraction of the watermark. In most SVD-based procedures, the watermark is embedded by controlling the singular values as per the watermarking bit. Authors (Wang et al. 2011) utilized diminished singular value decomposition technique which utilizes the unitary matrix for watermark bit inserting. A portion of the watermarking systems additionally exists that utilized the blend of DWT and SVD for watermark inserting and extraction.

The issue with the watermarking algorithm utilizing the SVD framework is that the SVD framework itself is straightforwardly inclined to the assault. There is no security key required through which the singular values, which are required to be utilized for watermark inserting are covered up. Since little change in the singular values doesn't influence the imperceptibility of the audio signal, the interloper can control a similar singular value of the SVD matrix. The prerequisite that, regardless of whether the algorithm of watermarking is known to the attacker, he ought not to have the access of the watermarking areas isn't met in the watermarking system utilizing SVD. The utilization of the singular values for watermark installing depends on the way that if there will be a slight change in the singular values it won't bother the straightforwardness of the image or audio and furthermore there is no unmistakable change in singular values when the image or audio is subjected to normal signal processing task. Along these lines, SVD-based audio watermarking scheme misuses this property to add the watermark data to the singular values of the diagonal matrix S or the sections of the unitary matrices such that inaudibility/imperceptibility isn't exasperated and robustness prerequisites of the powerful digital audio watermarking algorithm are accomplished. The SVD-based strategy varies in the diverse SV's utilization and the system through which the embedding is finished utilizing SVs (Wang et al. 2011).

3.5 Direct Sequence Spread Spectrum (DSSS) Encryption Algorithm

If the watermark embedding steps known in advanced, then watermark can be extracted easily by attackers. In any case, it is vital that the watermark is encoded before it is being embedded by which it will turn out to be about incomprehensible for the intruders to evacuate the watermark. Another vital thing in watermark embedding is that the vitality of the watermark is equitably disseminated all through the host signal. Else, the embedded signal appears as though it has more commotion implanted in it. For this purpose, the well-known algorithm is used in this research work that is known as DSSS (direct sequence spread spectrum).

Spread spectrum watermarking technique is a case of the correlation technique which inserts pseudorandom grouping and recognizes watermark by figuring the correlation between pseudorandom noise sequence and watermarked audio signal. Spread spectrum methods for watermarking get most of the hypothesis from the communication network (Czerwinski et al. 2000). The primary thought is to implant

Fig. 3.6 DSSS encoding method

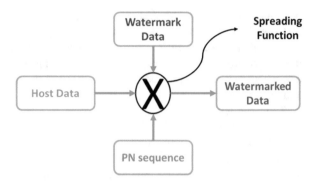

a limited band signal (the watermark) into a wide-band channel (the sound file). The qualities of both audio signal and watermark appear to suit the model superbly. In the expansion, spread spectrum strategies offer the plausibility of ensuring the watermark security by utilizing a secret key to control the pseudorandom sequence generator. Spread spectrum procedures permit the frequency bands to be coordinated previously embedding the message. This is the reason spread spectrum methods are significant not just for robustness correspondence yet to watermarking also. There are mainly two approaches for this spread spectrum technique, one is frequency hopping and another one is DSSS (direct sequence spread spectrum). The main objective of this method is to spread the watermark data onto the entire audio data which have the large frequency band.

In DSSS technique, a watermark data signal is embedded directly by the pseudo-random noise sequence (PN), which infers that a key is needed to encode and decode the bits (Peng et al. 2013). This DSSS encoding scheme is depicted in Fig. 3.6.

In this section, we gave a brief introduction of various techniques in the spatial domain as well as the transform domain. We also explained the working strategies of different transformation techniques, for example, DFT, DCT, and DWT. We likewise examined the noteworthiness of wavelet and its prevalence over other frequency domain methods.

References

Ali Khayam S (2003) The discrete cosine transform (DCT): theory and application, Information theory and coding, seminar 1—The discrete cosine transform: theory and application, March 10, 2003

Arnold M, Schmucker M, Wolthusen SD (2003) Techniques and applications of digital watermarking and content protection. Artech House, Inc., Boston, London

Bassia P, Nikolaidis N (2001) Robust audio watermarking in the time domain. IEEE Trans Multimedia 3(2):232–241

Bender W, Gruhl D, Morimoto N, Lu A (1996) Techniques for data hiding. IBM Syst J 35 (3–4):1996

Bhat V et al (2011) A new audio watermarking scheme based on singular value decomposition and quantization. Circuits Syst Signal Process 30:915–927

Chitode JS, Patil M (2013) Improved technique for audio watermarking based on discrete wavelet transform. Int J Eng Adv Technol 2(5):511

Czerwinski S, Fromm R, Hodes T (2000) Digital music distribution and audio watermarking (IS 219). University of California, Berkeley

FFT window functions: limits on FFT analysis (2009) Bores signal processing. http://www.bores.com/courses/advanced/windows/files/windows.pdf. Accessed 25 Oct 2009

Gopalan K (2003) Audio steganography using bit modification, Proceedings of the 2003 IEEE international conference on acoustics, speech & signal processing, China

Katzenbeisser S, Petitcolas FAP (2000) Information hiding techniques for steganography and digital watermarking. Artech House, Inc., Boston, London

Kim HJ, Choi YH, Seok J, Hong J (2003) Audio watermarking technique: intelligent watermarking techniques: theory and applications. World Scientific Publishing, Singapore

Mistry D (2010) Comparison of digital water marking methods. Int J Comput Sci Eng 2 (9):2905–2909

Özer H, Sankur B (2005) An SVD based audio watermarking technique. In: Proceedings of the IEEE 13th conference on signal processing and communications applications. IEEE, New York, pp 452–455

Peng H, Li B, Luo X, Wang J, Zhang Z (2013) A learning-based audio watermarking scheme using kernel Fisher discriminant analysis. J Digit Signal Process 23:382–389

Polikar R (2010) Home page—Dr. Robi Polikar, Jan 2001. [Online]. http://users.rowan.edu/~polikar/WAVELETS/WTtutorial.html. Accessed 21 July 2010

Potdar VM, Han S, Chang E (2005) A survey of digital image watermarking techniques, 0-7803-9094-6/05/$20.00 ©2005 IEEE

Rathee N, Kumar S (2014) A survey on audio watermarking techniques. Int J Eng Appl Manag Sci Paradigm 16(1)

Shahadi HI, Jidin R, Way WH, Abbas YA (2014) Efficient FPGA architecture for dual mode integer haar lifting wavelet transform core. J Appl Sci 14:436–444

Singh V (2011) Digital watermarking: a tutorial, Cyber Journals: Multidisciplinary Journals in Science and Technology, Journal of Selected Areas in Telecommunications (JSAT) January Edition

Singh J, Garg P, Nath A (2009) Audio watermarking using spectral modifications. Int J Signal Process 5(4):2009

Sun X, Liu J, Sun J, Zhang Q, Ji W (2008) A robust image watermarking scheme based-on the relationship of SVD. In: Proceedings of the international conference on intelligent information hiding and multimedia signal processing. IEEE, New York

Tan W, Yang S, Chen Y, Zhou J (2009) Research on DFT domain digital audio watermarking algorithm based on quantization', First international workshop on education technology and computer science, China

Wang X, Zhao H (2006) A novel synchronization invariant audio watermarking scheme based on DWT and DCT. IEEE Trans Signal Process 54(12):4835–4840

Wang J, Healy R, Timoney J (2011) A robust audio watermarking scheme based on reduced singular value decomposition and distortion removal. J Signal Process 91(8):1693–1708

Watson AB (1994) Image compression using the discrete cosine transform. Mathemat J 4(1):81–88

Xiaojuan X et al (2007) DWT-based audio watermarking using support vector regression and subsampling. In: Proceedings of the 7th international workshop on fuzzy logic and applications. IEEE, New York, pp 136–144

Xuehua J (2010) Digital watermarking and its application in image copyright protection. In: 2010 international conference on intelligent computation technology and automation. IEEE, New York

Yan Y, Rong H, Mintao X (2009) A novel audio watermarking algorithm for copyright protection based on DCT domain, Second international symposium on electronic commerce and security, China

Zhang X, Li K (2005) Comments on "An SVD-based watermarking scheme for protecting rightful ownership". IEEE Trans Multimedia 7(2):593–594

Chapter 4
Main Objective

Abstract In this chapter, the problem statement is identified by exploring and studying various existing methods. In this research, the main objective is to provide the robust and secured watermarking algorithm which can stand against the various malicious attacks by the attackers. And also, it maintains the quality of the original audio data as well as the watermark data.

Keywords Main objective · Robustness · Security

4.1 Identified Issues

Based on the writing survey, it can be expressed that the fundamental issue with the audio watermarking and with all the watermarking algorithms which utilize other sorts of cover protest is to make the watermarked question (which is inserted with additional data) vigorous to assaults while keeping up the imperceptibility. This issue turns out to be more genuine if there should arise an occurrence of audio watermarking in light of the sensitivity of HAS (human auditory system). The necessities of the audio watermarking repudiate with each different as robustness requires the watermark to be implanted in the conspicuous segment of the audio so it can't be evacuated through assaults. Be that as it may, this will diminish the imperceptibility. In this manner, an ideal exchange off is required to be kept up for imperceptibility, robustness, and security for the watermarking plans utilizing better embedding methodologies and subsequently, it is as yet an open issue. Some extra issues are distinguished which are as per the following:

Issue 1: Despite the fact that the DFT and DCT-based watermarking algorithms have low inserting unpredictability, however, the utilization of low frequency as the watermarking areas prompt less imperceptibility. There is a need to offer thoughtfulness regarding the utilization of chose frequency coefficients and better embedding methodology to give a decent harmony between imperceptibility and robustness. Instead of embedding watermark data into a single coefficient if that is to be embedded into the group of coefficient then it would provide the high robustness against the malicious attacks.

K. N. Patel, *Robust and Secured Digital Audio Watermarking*, SpringerBriefs in Speech Technology, https://doi.org/10.1007/978-3-030-53911-5_4

Issue 2: The reliability of the recovered watermark is a major concern. So, there is a need to offer thoughtfulness regarding the utilization of better embedding methodology to give a reliable solution.

Issue 3: The another major concern is security. Watermarking algorithms which are discussed in the literature survey using the images, text messages, and other media type data are primarily utilized as a watermark. There can be a circumstance when the watermark itself can be utilized to deceive the ownership. The watermark utilized by one can be utilized by numerous more people additionally and asserting the copyright on such watermark and eventually on the cover media turns out to be extremely troublesome. For this, we require the secure watermarking algorithm which provides better security against the unauthorized access.

4.2   Objectives of Research

In light of the writing survey done and the issues recognized alongside the primary issue of watermarking, the main objective of this book is arranged for imperceptibility, robustness, and security. The proposed audio watermarking algorithm is source based on, that is, proprietorship identification. The DWT, SVD, and DFT transform is utilized for embedding as they are all around acknowledged in the watermarking domain. The robustness against the basic signal processing assaults is a must as these audio files are easily and frequently provided from the internet. The module to produce a one of a kind watermark for proprietor validation and security against the likewise is by all accounts a most extreme prerequisite.

Based on this the objectives are defined as follows:

Objective 1: The first objective of this research is to implement the robust audio watermarking algorithm using DWT transformation method by choosing the different level of decomposition which will provide the high robustness against various malicious attacks and good imperceptibility.

Objective 2: The second objective is to develop the improved audio watermarking algorithm to minimize the degradation ration of original content ineffective way.

Objective 3: The third objective is to develop the efficient audio watermarking algorithm that would provide the best SNR value for the provided attacks for all types of audio data.

Objective 4: The fourth objective is to implement an improved audio watermarking algorithm using SVD is robust against various malicious attacks.

Objective 5: The fifth objective is to implement a secure and robust audio watermarking algorithm to provide the security using DSSS encryption method against unauthorized access and which will be used in different communication application like in defense system.

Chapter 5
Technique for Enhancing the Imperceptibility and Robustness of Digital Audio Watermarking

Abstract After discussing all these, our main research discussion starts from this chapter, in that digital audio watermarking algorithm is discussed. Initially, based on the survey done, the various methods available for digital audio watermarking are taken for consideration. From those methods, we found one approach that is relatively better than other approaches are proposed. It is based on the combination of DWT and SVD methods. So, we have implemented that algorithm first and found that it is better but still there are some improvements needed in that to increase the robustness because we found that this algorithm is not fitted for some kind of attacks as well some different types of watermark images. After that, we compare our results with existing methods and we found better results than other methods. But still, some improvements are needed in our work like in echo attack it is not giving the desired result. So, again we proposed another new approach that is the combination of DWT and SVD with DFT to increase the robustness and DSSS algorithm to provide the security to our algorithm. Our results show tremendous improvement against various malicious attacks and also improved the robustness.

Keywords Robustness · Security · Imperceptibility · DWT · SVD · DFT · DSSS · Watermark embedding · Watermark extraction · Various attacks

In this proposed approach, two most authoritative methods are used: discrete wavelet transform (DWT) and singular vector decomposition (SVD) to improve the imperceptibility and robustness. Also for the security purpose, the 16-bit synchronized key is used for encryption and decryption of the watermark. Here, two algorithms are represented for watermark embedding and watermark extracting along with block diagrams of digital audio watermarking technique. In the journey of our research work, two improved algorithms have been proposed for the digital audio watermarking scheme.

- Proposed Method—1

In this proposed method, two most powerful methods are used: discrete wavelet transform (DWT) and singular vector decomposition (SVD) to improve the imperceptibility and robustness and the synchronized secret key is used to provide the security. Here, two algorithms are represented for watermark embedding and watermark extracting along with block diagrams of audio watermarking technique.

5.1   Audio Watermark Embedding Algorithm Using DWT-SVD

The watermark embedding procedure transforms the audio signal using DWT and SVD, embeds the bits of a binary image watermark in appropriate locations in the transformed signal, and finally produces a watermarked audio signal by performing inverse SVD and DWT operations. The procedure is illustrated in the block diagram shown in Fig. 5.1 and described thereafter.

1. According to the block diagram shown in Fig. 5.1. Read the original audio file and do the block structuring to store the watermark image bits for embedding process. After that store, the values of right and left side channels in particular variables say $(m \times n)$.

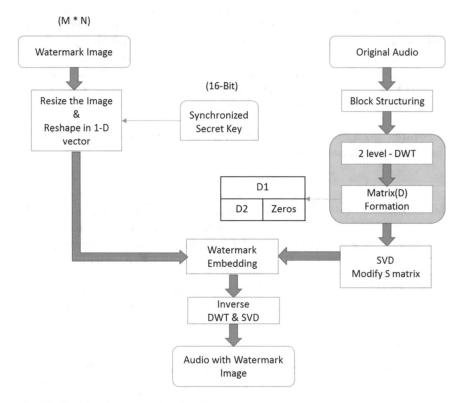

Fig. 5.1 Digital audio watermark embedding process

Fig. 5.2 Two-level DWT

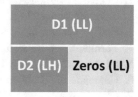

Fig. 5.3 Formulation of
D matrix

2. After that check whether the watermark image size, here we say it ($W_{M \times N}$), is satisfying the following criteria for the embedding process: (As this is generalizing algorithm, here we can embed audio or text file as well instead of watermark image)

$$m \times n \geq M \times N \qquad (5.1)$$

where m, n are the audio size and M, N is the watermark image size variables.
3. Here to provide the security watermark image is encrypted by 16-bit shared synchronized secret key $[Cn]$ in the time domain.
4. In the next step, reconstruction of watermark image is to be done into ($1 \times MN$) size for embedding process, because we want to embed the bits into alternative fashion in left and right channels of the audio file which are in 1-D form.
5. Obtain the two-level DWT to get the lowest frequency component in the LL segment as shown in Fig. 5.2.

6. In the watermark embedding process lowest frequency component is required, for that signal is decomposed for 2 level DWT and after getting that D matrix is reconstructed to find the SVD in Fig. 5.3.

7. Based on the calculated D matrix, we find the singular vector decomposition and modify only S_{11} component using the following formula (do not modify the U and V matrix):

$$S_{11\,new} = S_{11\,old} + \alpha * W_n \qquad (5.2)$$

where α = watermark intensity and W_n = nth bit of watermark image, S_{11new} is watermark first bit (1 row, 1 col)
8. Now, find the inverse singular vector decomposition using the newly obtained value of S_{11new} by following:

$$D = U * S_{11\,\text{new}} * V^T \qquad (5.3)$$

9. In last, find the inverse DWT and SVD to get the original audio with the watermarked embedded image.

The following are the watermark embedding algorithm steps:

1. Take the original Audio file ($m \times n$).
2. Take the image ($M \times N$) to be watermarked and reshape in ($1 \times MN$).
3. Embed synchronized secret key with watermark image (time domain).
4. Formation of the audio file in desired blocks for watermark embedding process.
5. Obtain two-level DWT and formation of D matrix.
6. Calculate the SVD components (U, $S_{11\text{old}}$, V^T).
7. Change the S_{11} bit of S component using formula $S_{11\text{new}} = S_{11\text{old}} + \alpha \times W_n$.
8. Find inverse SVD using the formula: $U \times S_{11}\text{new} \times V^T$ and find inverse DWT.

5.2   Audio Watermark Extracting Algorithm Using DWT-SVD

1. According to the block diagram shown in Fig. 5.4. Read the watermarked audio file and do the block structuring to store the watermark image bits for the extraction process. After that store, the values of right and left side channels in particular variables say ($m \times n$).
2. In the next step, reconstruction of watermark image is to be done into ($1 \times MN$) size for embedding process, because we want to embed the bits into alternative fashion in left and right channels of the audio file which are in 1-D form.
3. Obtain the two-level DWT to get the lowest frequency component in the LL segment as shown in Fig. 5.2.
4. Based on the obtained DWT reconstruct the D matrix as shown in Fig. 5.3.
5. Based on the calculated D matrix, we find the singular vector decomposition and modify only S_{11} component using Eq. (5.2) (do not modify the U and V matrix).
6. Now, find watermarked image bit (W_n) based on the calculated $S_{11\text{new}}$ using formula:

$$W_n = 1\{S_{11\,\text{old}}/S_{11\,\text{new}} \geq 1\}$$
$$W_n = 0\{S_{11\,\text{old}}/S_{11\,\text{new}} \leq 0\} \qquad (5.4)$$

7. Finally, Extract watermark image using Shared synchronized secret key.

The following are the watermark extraction algorithm steps:

1. Take the watermarked audio file ($m \times n$).
2. Formation of the audio file in desired blocks for watermark extraction process.

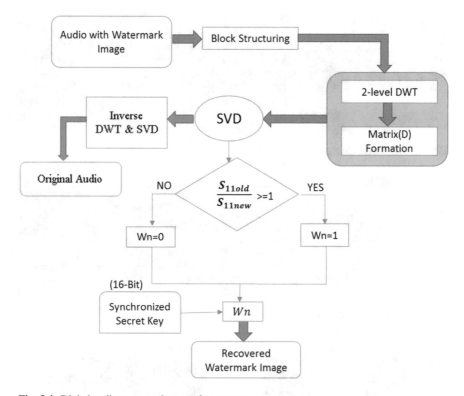

Fig. 5.4 Digital audio watermark extracting process

3. Obtain two-level DWT and formation of D matrix.
4. Calculate the SVD components (U, S_{11new}, V).
5. Find watermarked image bit (W_n) based on a calculated S_{11new} using formula.
   If ($S_{11old}/S_{11new} \geq 1$)
   $$W_n = 1$$
   Else
   $$W_n = 0$$
   End
6. Extract watermark image using Shared synchronized secret key.

5.3   Results for Digital Audio Watermarking (Proposed Method—1)

In this research, we have used different music type audio files (.wav and .mp3) and images. From which we have used four different music types of audio files as cover audio. We have tried different attacks on the cover audio as well as watermark image to validate the proposed method.

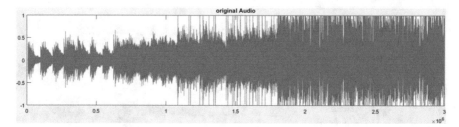

Fig. 5.5 Original audio signal

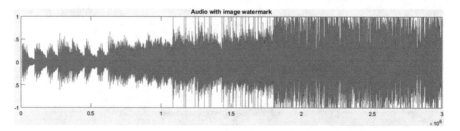

Fig. 5.6 Audio signal with image watermark

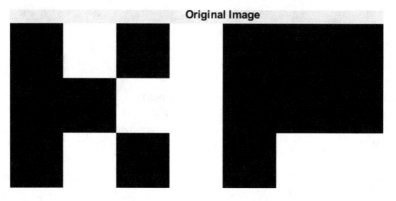

Fig. 5.7 Original watermark image

By applying the proposed algorithm, the following results are coming up. That shows the comparatively good results than the previously proposed algorithms. The following results were obtained by applying various attacks on the watermarked audio file (Figs. 5.5, 5.6, 5.7, 5.8, 5.9, 5.10, 5.11, 5.12, 5.13, 5.14, 5.15, and 5.16; Tables 5.1, 5.2, 5.3, 5.4, 5.5, 5.6, 5.7, and 5.8).

To approve the robustness of cover audio file and perceptual transparency of watermark image we have applied distinctive assaults/attacks on the watermarked image acquired from proposed strategy, that is, digital audio watermarking using DWT and SVD.

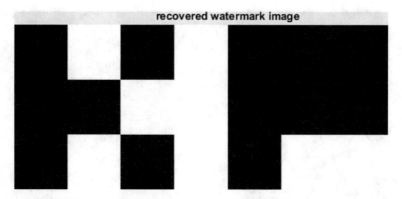

Fig. 5.8 Recovered watermark image

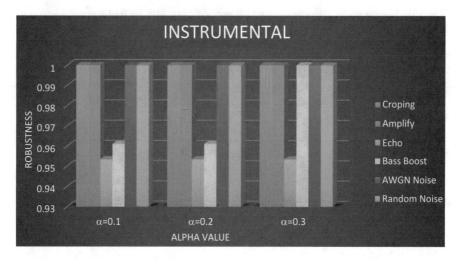

Fig. 5.9 Comparison of different intensity values to calculate robustness for (instrumental music type)

Furthermore, in this research, we have applied amplifying, AWGN noise, random noise, cropping, and echo attacks on the watermarked image. Different tables are prepared for the better comparison of different robustness values and SNR values calculated for the proposed method. As we can see from the above tables values that we are getting different values for the calculated robustness and SNR values for different music types with provided intensity values. From that, we can get the results below.

From that, we can tell that for some types of attacks still, we had not got the desired results for the proposed algorithm like the echo attack. But still, we are able to achieve better results in comparison with the previous methods proposed by different authors. Below figure shows the results of some values (Fig. 5.17).

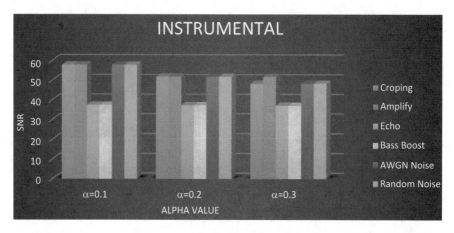

Fig. 5.10 Comparison of different intensity values to calculate SNR for (instrumental music type)

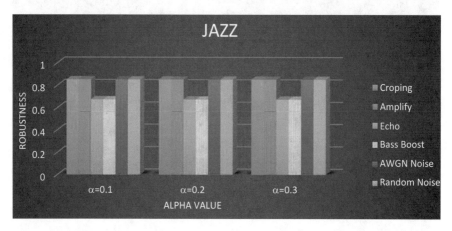

Fig. 5.11 Comparison of different intensity values to calculate robustness for (jazz music type)

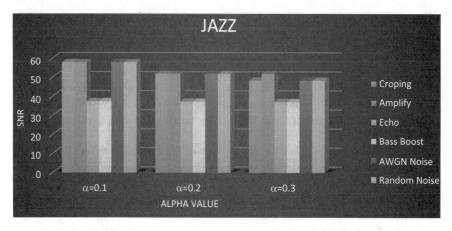

Fig. 5.12 Comparison of different intensity values to calculate SNR for (jazz music type)

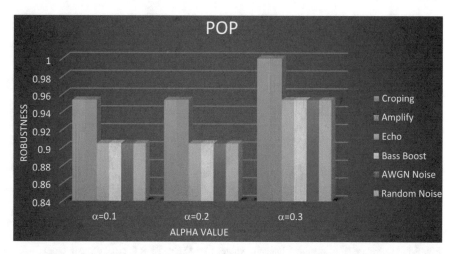

Fig. 5.13 Comparison of different intensity values to calculate robustness for (pop music type)

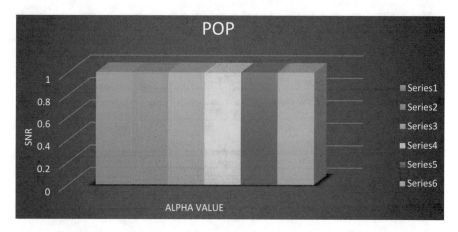

Fig. 5.14 Comparison of different intensity values to calculate SNR for (pop music type)

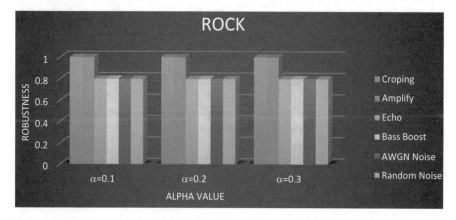

Fig. 5.15 Comparison of different intensity values to calculate robustness for (rock music type)

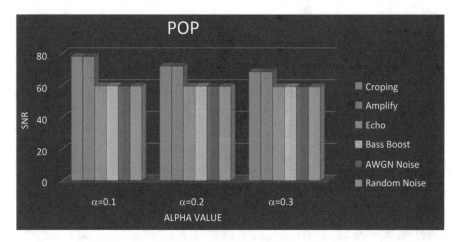

Fig. 5.16 Comparison of different intensity values to calculate SNR for (rock music type)

Table 5.1 Calculated values for robustness after applying different attacks for (instrumental music type)

Music type	Attacks	Robustness		
		$\alpha = 0.1$	$\alpha = 0.2$	$\alpha = 0.3$
Instrumental	Cropping	1	1	1
	Amplify	1	1	1
	Echo	0.9534	0.9534	0.9534
	Bass boost	0.9611	0.9611	1
	AWGN noise	1	1	1
	Random noise	1	1	1

Table 5.2 Calculated values for SNR after applying different attacks for (instrumental music type)

Music type	Attacks	SNR		
		$\alpha = 0.1$	$\alpha = 0.2$	$\alpha = 0.3$
Instrumental	Cropping	65.2410	65.6212	66.9580
	Amplify	65.2410	65.6212	66.9580
	Echo	65.2223	65.4317	66.2210
	Bass boost	65.2397	65.4908	66.5321
	AWGN noise	65.2410	65.6212	66.9580
	Random noise	65.2410	65.6212	66.9580

Below is the following table the comparison of the proposed method with the previous work is depicted (Fig. 5.18; Table 5.9).

Table 5.3 Calculated values for robustness after applying different attacks for (jazz music type)

Music type	Attacks	Robustness		
		$\alpha = 0.1$	$\alpha = 0.2$	$\alpha = 0.3$
JAzz	Cropping	0.8528	0.8528	0.8528
	Amplify	0.8528	0.8528	0.8528
	Echo	0.6741	0.6741	0.6741
	Bass boost	0.6741	0.6741	0.6741
	AWGN noise	0.8528	0.8528	0.8528
	Random noise	0.8528	0.8528	0.8528

Table 5.4 Calculated values for SNR after applying different attacks for (jazz music type)

Music type	Attacks	SNR		
		$\alpha = 0.1$	$\alpha = 0.2$	$\alpha = 0.3$
JAzz	Cropping	57.8573	51.8367	48.3149
	Amplify	57.8573	51.8367	51.8367
	Echo	37.4000	37.2737	37.0737
	Bass boost	37.4000	37.2737	37.0737
	AWGN noise	57.8573	51.8367	48.3149
	Random noise	57.8573	51.8367	48.3149

Table 5.5 Calculated values for robustness after applying different attacks for (pop music type)

Music type	Attacks	Robustness		
		$\alpha = 0.1$	$\alpha = 0.2$	$\alpha = 0.3$
POP	Cropping	0.9534	0.9534	1
	Amplify	0.9534	0.9534	1
	Echo	0.9045	0.9045	0.9534
	Bass boost	0.9045	0.9045	0.9534
	AWGN noise	0.9045	0.9045	0.9534
	Random noise	0.9045	0.9045	0.9534

Table 5.6 Calculated values for SNR after applying different attacks for (pop music type)

Music type	Attacks	SNR		
		$\alpha = 0.1$	$\alpha = 0.2$	$\alpha = 0.3$
POP	Cropping	78.0438	72.0232	68.5014
	Amplify	78.0438	72.0232	68.5014
	Echo	59.5296	59.3535	59.0728
	Bass boost	59.5296	59.3535	59.0728
	AWGN noise	59.5296	59.3535	59.0728
	Random noise	59.5296	59.3535	59.0728

Table 5.7 Calculated values for robustness after applying different attacks for (rock music type)

Music type	Attacks	Robustness		
		$\alpha = 0.1$	$\alpha = 0.2$	$\alpha = 0.3$
ROCK	Cropping	1	1	1
	Amplify	1	1	1
	Echo	0.7977	0.7977	0.7977
	Bass boost	0.7977	0.7977	0.7977
	AWGN noise	0.7977	0.7977	0.7977
	Random noise	0.7977	0.7977	0.7977

Table 5.8 Calculated values for SNR after applying different attacks for (rock music type)

Music type	Attacks	SNR		
		$\alpha = 0.1$	$\alpha = 0.2$	$\alpha = 0.3$
POP	Cropping	60.2755	54.2549	50.7331
	Amplify	60.2755	54.2549	50.7331
	Echo	40.9665	40.9665	40.9665
	Bass boost	40.9665	40.3866	40.3866
	AWGN noise	40.9665	40.3866	40.3866
	Random noise	40.9665	40.3866	40.3866

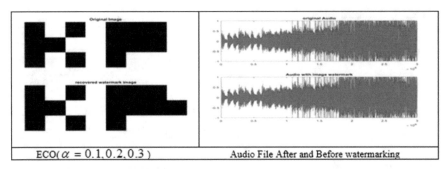

ECO($\alpha = 0.1, 0.2, 0.3$)	Audio File After and Before watermarking

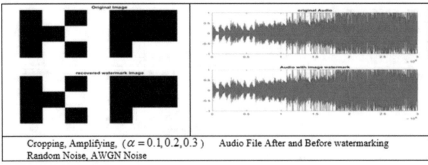

Cropping, Amplifying, ($\alpha = 0.1, 0.2, 0.3$) Random Noise, AWGN Noise	Audio File After and Before watermarking

Fig. 5.17 Recovered and original watermark images and audio file signals

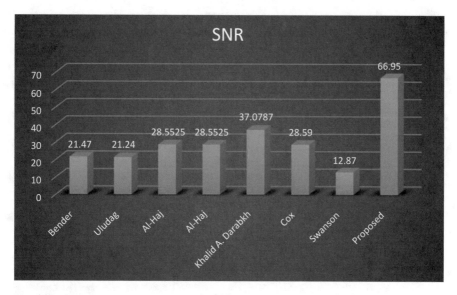

Fig. 5.18   Comparison of SNR values with previous work

Table 5.9   Comparison of proposed algorithm SNR with previous work

Author	Method	SNR
Bender et al. (1996)	Echo	21.47
Uludag and Arslan (2001)	DC-level shifting	21.24
Al-Haj and Mohammad (2010)	DWT-SVD	38.17
Al-Haj et al. (2011)	DWT	28.5525
Darabkh (2014)	DWT-SVD	37.0787
Cox et al. (1997)	Spread spectrum	28.59
Swanson et al. (1998)	Frequency masking	12.87
Proposed	DWT-SVD	66.95

5.4   Summary

From the results tested against various attacks, it is observed that by using the proposed algorithm we have achieved maximum robustness and imperceptibility with security. We have improved the results compared to previous work done in this field. By this data, hiding becomes more robust and original audio, as well as watermark image, can be easily recovered with minimal effort and it is also imperceptible against various attacks and distortion. It is also observed that for the attacks like echo and bass boost the results are more than good though this is hard to achieve, the proposed algorithm gives better results for this. Future work will be focused as enhancing the proposed method against more attacks and alterations.

References

Al-Haj A, Mohammad A (2010) Digital audio watermarking based on the discrete wavelets transform and singular value decomposition. Eur J Sci Res 39:6–21

Al-Haj A, Mohammad A, Bata L (2011) DWT-based audio watermarking. Int Arab J Informat Technol 8:326–333

Bender W, Gruhl D, Morimoto N, Lu A (1996) Techniques for data hiding. IBM Syst J 35 (3–4):1996

Cox IJ, Kilian J, Leighton T, Shamoon T (1997) Secure spread spectrum watermarking for multimedia. IEEE Trans Image Process 6:1673–1687

Darabkh KA (2014) Imperceptible and robust DWT-SVD-based digital audio watermarking algorithm. J Softw Eng Appl 7:859–871. https://doi.org/10.4236/jsea.2014.710077

Swanson MD, Zhu B, Tewfik AH, Boney L (1998) Robust audio watermarking using perceptual masking. Signal Process 66:337–355. https://doi.org/10.1016/S0165-1684(98)00014-0

Uludag U, Arslan L (2001) Audio watermarking using DC-level shifting. Project Report. Bogazici University, Istanbul. http://www.busim.ee.boun.edu.tr/speechweb/

Chapter 6
Technique for Enhancing the Robustness, Imperceptibility, and the Security of Digital Audio Watermarking

Abstract This chapter describes the results produced by the new algorithm and also performance analysis by comparing all the results with the existing algorithms has been discussed.

Keywords Results · Comparison

For the improvement of the algorithm, we have increased the DWT decomposing levels and we have made some modification in the algorithm as follows:

- **Proposed Method—2**

In this proposed method, three most adopted methods are used: discrete wavelet transform (DWT), singular vector decomposition (SVD), and discrete Fourier transform (DFT) to improve the imperceptibility and robustness and direct sequence spread spectrum (DSSS) encryption algorithm is used to provide the security for the digital audio watermarking algorithm. Here, two algorithms are represented for watermark embedding and watermark extracting along with block diagrams of audio watermarking technique.

6.1 Audio Watermark Embedding Algorithm Using DWT-SVD-DFT

The watermark embedding procedure transforms the audio signal using DWT and SVD, embeds the bits of a binary image watermark in appropriate locations in the transformed signal, and finally produces a watermarked audio signal by performing inverse SVD and DWT operations. The procedure is illustrated in the block diagram shown in Fig. 6.1 and described thereafter.

1. Change over the double picture watermark into a one-dimensional vector b of length $m \times n$. A watermark bit B_i may take one of two qualities: 0 or 1.

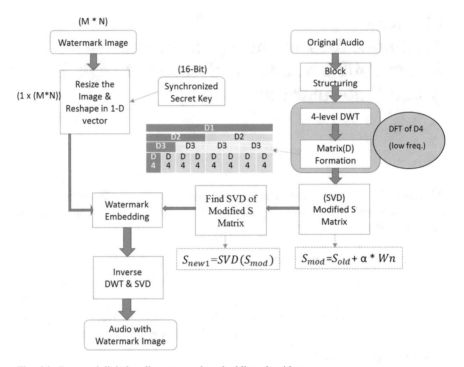

Fig. 6.1 Proposed digital audio watermark embedding algorithm

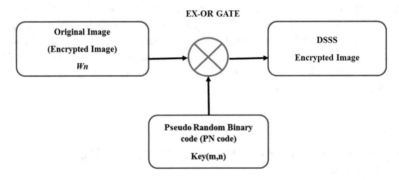

Fig. 6.2 DSSS encryption process

$$B_i = \{[0, 1], 1 \le i \le (m \times n)\} \tag{6.1}$$

2. We have to encrypt the message image or original image with pseudorandom (PN) code/key generated at the time of embedding using DSSS system (Fig. 6.2).

Fig. 6.3 Four-level DWT

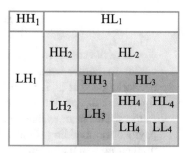

Fig. 6.4 Formulation of the *D* matrix

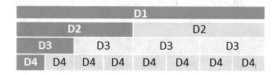

$$\text{Msg\_Image} = \text{EXOR}\,(\text{bi}, \text{PN}) \tag{6.2}$$

3. Sample the first audio signal at a testing rate of 44,100 examples for every second and divide the inspected record into N frames. The best or more favorable frame length will be determined by means of experiment in such a way to rise data load.
4. Perform a four-level DWT transformation on each frame. This operation produces five multi-resolution sub-bands: D_1, D_2, D_3, D_4, and A_4. The D sub-bands are called "detail sub-bands" and the A_4 sub-band is called "approximation sub-band." The five sub-bands are arranged in the vector shown in Fig. 5.8. Find the lowest frequency component of D_4 matrix using fast Fourier transformation (DFT).
5. Obtain the four-level DWT to get the lowest frequency component in the LL segment as shown in Fig. 6.3.

6. Arrange the four detail sub-groups D_1, D_2, D_3, and D_4 in a network D as appeared in Fig. 6.4. The lattice development is done along these lines to appropriate the watermark bits all through the multi-determination sub-groups D_1, D_2, D_3, and D_4. Shaping the network with the D_n, instead of utilizing an alone, is done to take into account grid development and subsequent utilization of the lattice-based SVD administrator. The extent of grid D is $4 \times (L/2)$, where L alludes to the length of the casing.

7. Using SVD decomposes the D_i matrix. That produces the three orthonormal grids Σ, U, and V^T as follows:

$$D_i = U_i * \sum *V_i^{\mathrm{T}} \tag{6.3}$$

where the corner to corner lattice Σ has a similar size of the D_i network. The corner to corner sections compares to the particular estimations of the D_i network. Be that as it may, for implanting purposes, just a (4 × 4) subset of framework Σ, appointed the name S from there on, is utilized as demonstrated as follows. This is an exchange off between imperceptibility (unintelligibility) and payload (embedding limit). That is, utilizing the entire Σ framework for implanting will increment installing limit, however, will prompt serious bending in imperceptibility (unintelligibility) of the watermarked audio signal.

$$K_{\mathrm{matrix}}(K_{\mathrm{tr}}) = \begin{vmatrix} K_{11} & 0 & 0 & 0 \\ 0 & K_{22} & 0 & 0 \\ 0 & 0 & K_{33} & 0 \\ 0 & 0 & 0 & K_{44} \end{vmatrix} \tag{6.4}$$

8. Arrange 12 bits of the first watermark bit vector b into a scaled 4 × 4 watermark matrix W. The watermark bits must be situated in the non-corner to corner positions inside the lattice, as demonstrated as follows.

$$W_n = \begin{vmatrix} 0 & b_0 & b_1 & b_2 \\ b_3 & 0 & b_4 & b_5 \\ b_6 & b_7 & 0 & b_8 \\ b_9 & b_{10} & b_{11} & 0 \end{vmatrix} \tag{6.5}$$

9. For instance, the watermark 12-bit watermark design 1010 0011 0101 must be changed over to the accompanying matrix frame before the real embedding is done.
10. Embed watermark grid W bits into framework K as per the accompanying "added substance implanting" equation:

$$K_{\mathrm{w}} = K_{\mathrm{tr}} + \alpha * W_n \tag{6.6}$$

where K_{w} is the watermarked K grid and α is the water-mark intensity which ought to be tuned the exchange off amongst robustness and imperceptibility. With this sort of implanting, the particular estimations of D_i stay unaltered, and therefore, capable of being heard mutilation caused by modifying the solitary qualities are dodged.
11. Decompose the new watermarked network K_{w} utilizing the SVD administrator. This task produces three new orthonormal matrices as takes after:

$$K_{\mathrm{w}} = U_{\mathrm{d}} * K_{\mathrm{new1}} * V_{\mathrm{d'}} \tag{6.7}$$

12. Make U_1 and V_1 vector for inverse SVD using.

$$U_1(:,:,i) = U_d \text{ and } V_1(:,:,i) = V_d \tag{6.8}$$

where $i = $ block size

13. Generate $K_{\_new}$ vector using K_{new1}

$$K_{\_new}(1:4,1:4) = K_{new1} \tag{6.9}$$

14. Apply the converse SVD activity utilizing the U, $K_{\_new}$, and V^T frameworks, which were unaltered, and the $K_{\_new}$ matrix, which has been adjusted by Eq. (6.9). The D matrix given underneath is the watermarked:

$$D_{wt} = U_w * K_{\_new} * V_w' \tag{6.10}$$

15. In last, find the inverse DWT to get the original audio with the watermarked embedded image.
16. Repeat every single past advance on each casing. The general watermarked audio signal is acquired by linking the watermarked outlines got in the past advances.

6.2   Audio Watermark Extracting Algorithm Using DWT-SVD-DFT

Given the watermarked audio signal and the relating U_1 and V_1 networks that were registered in Eq. (6.8) and put away for each casing, the installed watermark can be extricated by the strategy laid out in Fig. 5.9 and depicted in detail in the take after following steps:

1. Perform a four-level DWT transformation on each frame. This operation produces five multi-resolution sub-bands: D_1, D_2, D_3, D_4, and A_4. The D sub-bands are called "detail sub-bands" and the A_4 sub-band is called "approximation sub-band." The five sub-bands are arranged in the vector shown in Fig. 6.4.
2. Obtain the four-level DWT to get the lowest frequency component in the LL segment as shown in Fig. 6.3.
3. Arrange the four detail sub-groups D_1, D_2, D_3, and D_4 in a network D as appeared in Fig. 6.4. The lattice development is done along these lines to appropriate the watermark bits all through the multi-determination sub-groups D_1, D_2, D_3, and D_4. Shaping the network with the D_n, instead of utilizing an alone, is done to take into account grid development and subsequent utilization of the lattice-based SVD administrator. The extent of grid D is $4 \times (L/2)$, where L alludes to the length of the casing.

4. Using SVD decomposes the D_i matrix. That produces the three orthonormal grids Σ, U, and V^{T} as follows:

$$D_i = U_i * \sum *V_i^{\mathrm{T}} \qquad (6.11)$$

where the corner to corner lattice Σ has a similar size of the D_i network. The corner to corner sections compares to the particular estimations of the D_i network. Be that as it may, for embedding purposes, just a (4 × 4) subset of framework Σ, appointed the name S from there on, is utilized as demonstrated as follows. This is an exchange off between imperceptibility (unintelligibility) and payload (embedding limit). That is, utilizing the entire Σ framework for embedding will increment installing limit, however, will prompt serious bending in imperceptibility (unintelligibility) of the watermarked audio signal.

5. Obtain the matrix $S_{\_tr}$ from S matrix derived from SVD decomposition of D Matrix Using the given equation:
   $K_{\_tr} = K(1:4, 1:4)$ and U_{1p}, V_{1p} (U_1, V_1 matrix from watermark embedding procedure using equation.

$$U_{1p} = U_1(:,:,i) \text{ and } V_{1p} = V_1(:,:,i) \qquad (6.12)$$

where i = block size

6. Find new K_{matrix} from the given equation:

$$K_{wp} = U_{1p} * K_{\_tr} * V_{1p'} \qquad (6.13)$$

7. Extract the 12 watermark bits from each edge by looking at the non-askew estimations of lattice K_{wp}. It has been tentatively seen that there are two gatherings of non-corner to corner esteems that are extremely unmistakable. The qualities at the positions where a 0 bit has been inserted have a tendency to be significantly littler than those qualities at the positions where a 1 bit has been implanted. Therefore, to decide the watermark bit $W(n)$, the normal of non-inclining esteems is first registered, name it avg., at that point for each non-corner to corner esteem K_{wpij}, $W(n)$ is separated by the accompanying recipe:

$$W_n = \begin{cases} 0 & K_{wij'} \leq avg \\ 1 & else \end{cases} \qquad (6.14)$$

$$avg = 0.1 * max(K_w) \qquad (6.15)$$

8. Lastly, we have to decrypt the recovered image with pseudo-random (PN) code/key generated at the time of embedding using DSSS system as Figs. 5.6 and 6.5.

$$Msg\_Image = EXOR(W_n, PN) \qquad (6.16)$$

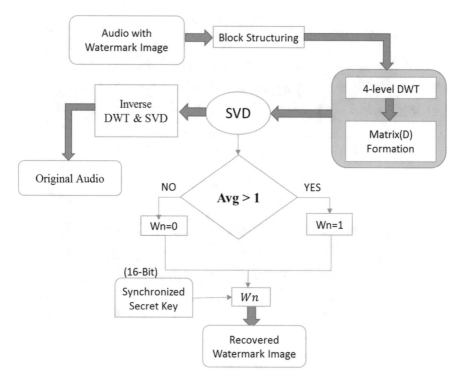

Fig. 6.5 Proposed digital audio watermark extracting algorithm

6.3   Results for Digital Audio Watermarking (Proposed Method—2)

As we have discussed earlier in the proposed method −1 for some attacks like echo attack and others we need to improve the algorithm for better robustness and imperceptibility for the audio file. Here, also we have checked the proposed algorithm for different music types and by applying various attacks on the watermarked audio file. In this proposed algorithm, DSSS encryption method is used to increase the security of the audio file (Figs. 6.6, 6.7, 6.8, 6.9, 6.10, 6.11, 6.12, 6.13, 6.14, 6.15, 6.16, 6.17, 6.18, 6.19, 6.20, 6.21, 6.22, 6.23, 6.24, 6.25, 6.26, 6.27, 6.28, 6.29, 6.30, 6.31, 6.32, and 6.33; Tables 6.1, 6.2, 6.3, 6.4, 6.5, 6.6, 6.7, 6.8, 6.9, 6.10, 6.11, 6.12, 6.13, 6.14, 6.15, 6.16, 6.17, 6.18, 6.19, 6.20, 6.21, 6.22, 6.23, 6.24, 6.25, 6.26, and 6.27).

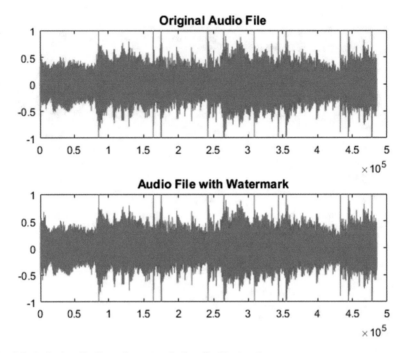

Fig. 6.6 Ordinal audio file and watermarked audio file signal

Fig. 6.7 Original image
and recovered watermark
image

Fig. 6.8 Original image
and recovered secured
encrypted watermark image

Encrypted Image

Recovered Encrypted Image

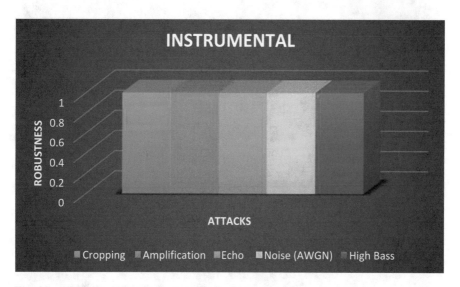

Fig. 6.9 Calculated values of robustness for (instrumental music type)

6.4   Summary

From the results tested against various attacks, it is observed that by using the
proposed algorithm we have achieved maximum robustness and imperceptibility
with security. We have improved the results compared to previous work done in this
field. By this data, hiding becomes more robust and original audio, as well as
watermark image, can be easily recovered with minimal effort and it is also imperceptible against various attacks and distortion. It is also observed that for the attacks
like echo and bass boost the results good though this is hard to achieve, the proposed
algorithm gives better results for this.

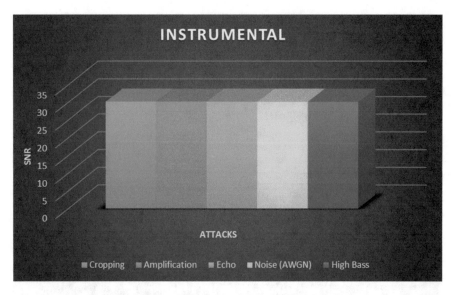

Fig. 6.10 Calculated values of SNR for (instrumental music type)

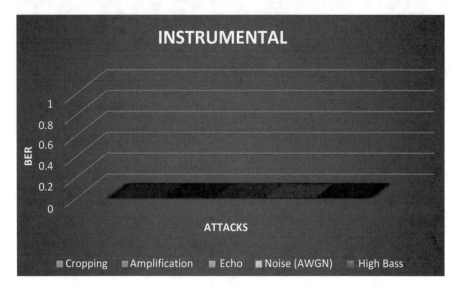

Fig. 6.11 Calculated values of BER for (instrumental music type)

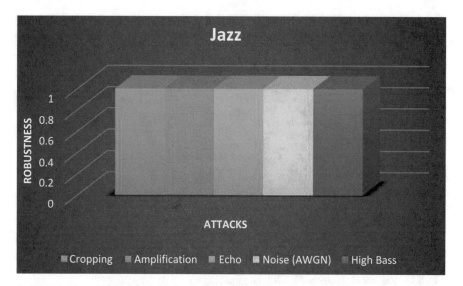

Fig. 6.12   Calculated values of robustness for (jazz music type)

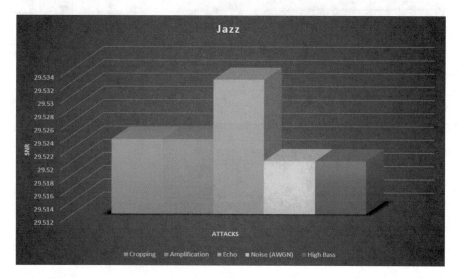

Fig. 6.13   Calculated values of SNR for (jazz music type)

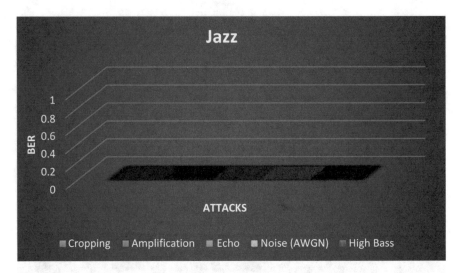

Fig. 6.14 Calculated values of BER for (jazz music type)

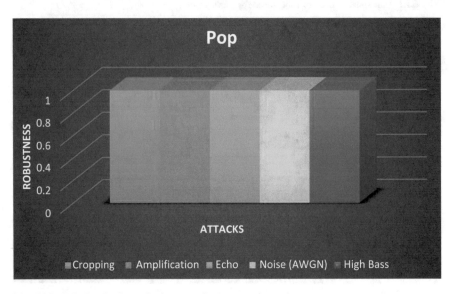

Fig. 6.15 Calculated values of robustness for (pop music type)

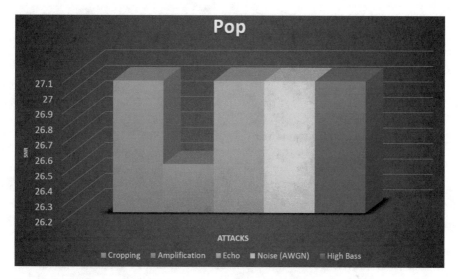

Fig. 6.16   Calculated values of SNR for (pop music type)

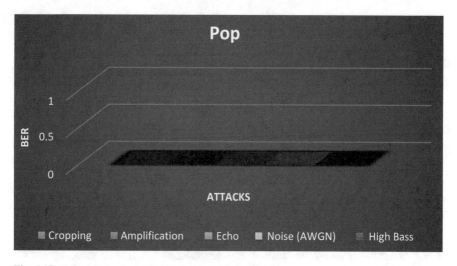

Fig. 6.17   Calculated values of BER for (pop music type)

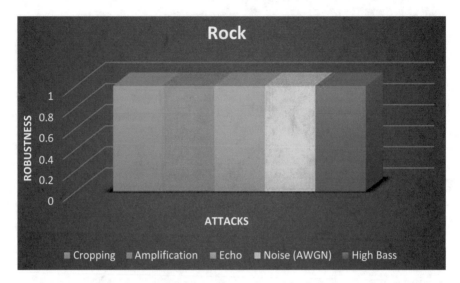

Fig. 6.18 Calculated values of robustness for (rock music type)

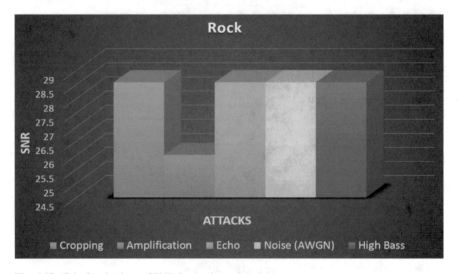

Fig. 6.19 Calculated values of SNR for (rock music type)

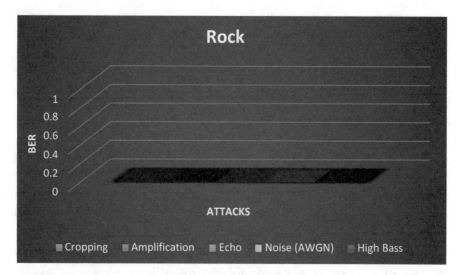

Fig. 6.20   Calculated values of BER for (rock music type)

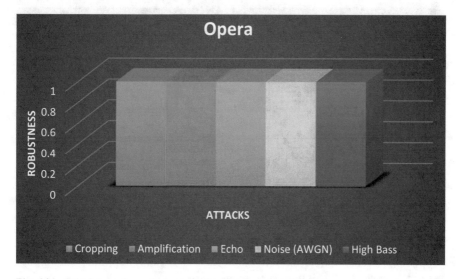

Fig. 6.21   Calculated values of robustness for (opera music type)

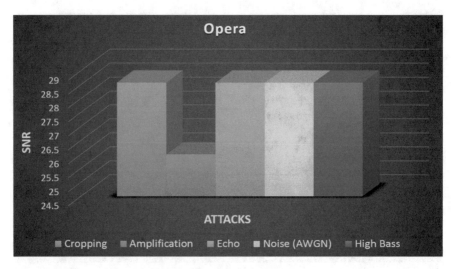

Fig. 6.22 Calculated values of SNR for (opera music type)

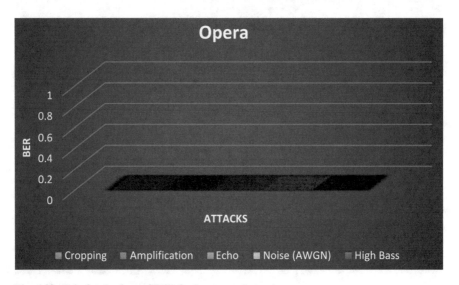

Fig. 6.23 Calculated values of BER for (opera music type)

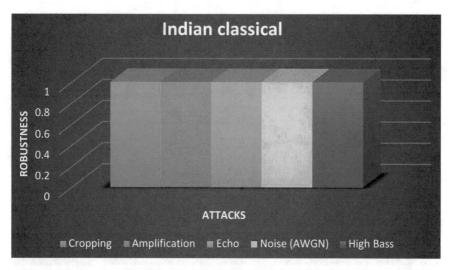

Fig. 6.24 Calculated values of robustness for (Indian classical music type)

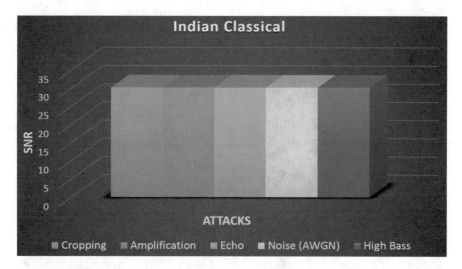

Fig. 6.25 Calculated values of SNR for (Indian classical music type)

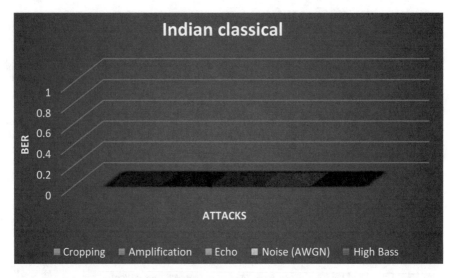

Fig. 6.26 Calculated values of BER for (Indian classical music type)

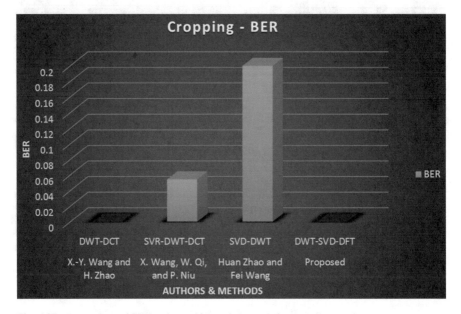

Fig. 6.27 Comparison of BER values with previous work for cropping attack

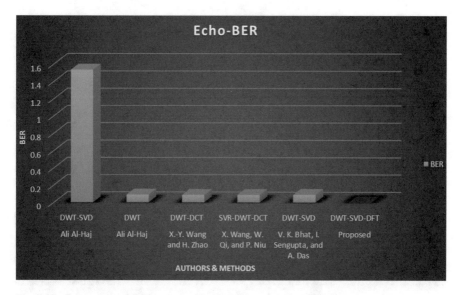

Fig. 6.28 Comparison of BER values with previous work for echo attack

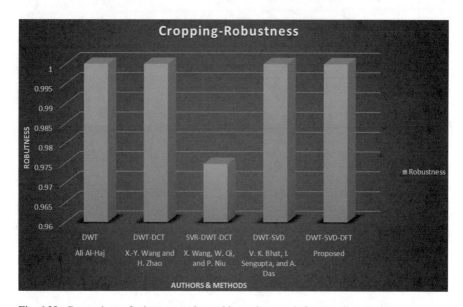

Fig. 6.29 Comparison of robustness values with previous work for cropping attack

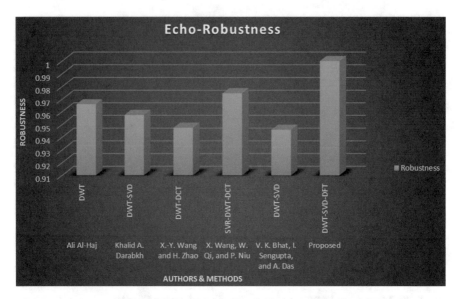

Fig. 6.30 Comparison of robustness values with previous work for echo attack

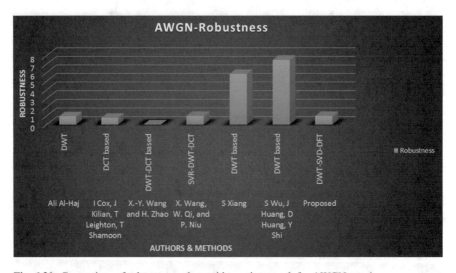

Fig. 6.31 Comparison of robustness values with previous work for AWGN attack

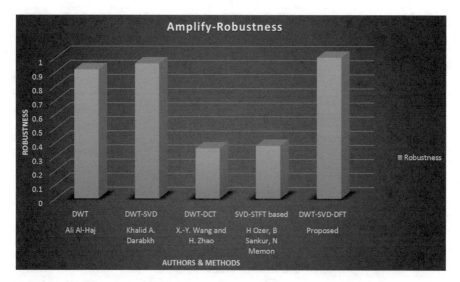

Fig. 6.32 Comparison of robustness values with previous work for amplify attack

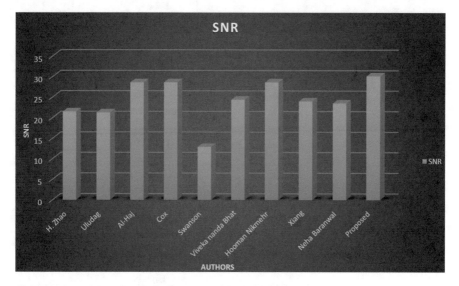

Fig. 6.33 Comparison of SNR values with previous work

Table 6.1 Calculated values of robustness after applying different attacks for (instrumental music type)	Music type	Attacks	Robustness
	Instrumental	Cropping	1
		Amplification	1
		Echo	1
		Noise (AWGN)	1
		High bass	1

Table 6.2 Calculated values of SNR after applying different attacks for (instrumental music type)

Music type	Attacks	SNR
Instrumental	Cropping	30.126
	Amplification	30.126
	Echo	30.126
	Noise (AWGN)	30.126
	High bass	30.126

Table 6.3 Calculated values of BER after applying different attacks for (instrumental music type)

Music type	Attacks	BER
Instrumental	Cropping	1
	Amplification	1
	Echo	1
	Noise (AWGN)	1
	High bass	1

Table 6.4 Calculated values of robustness after applying different attacks for (jazz music type)

Music type	Attacks	Robustness
Jazz	Cropping	1
	Amplification	1
	Echo	1
	Noise (AWGN)	1
	High bass	1

Table 6.5 Calculated values of SNR after applying different attacks for (jazz music type)

Music type	Attacks	SNR
Jazz	Cropping	29.5233
	Amplification	29.5233
	Echo	29.5323
	Noise (AWGN)	29.5200
	High bass	29.5200

Table 6.6 Calculated values of BER after applying different attacks for (jazz music type)

Music type	Attacks	BER
Jazz	Cropping	1
	Amplification	1
	Echo	1
	Noise (AWGN)	1
	High bass	1

Table 6.7 Calculated values of robustness after applying different attacks for (pop music type)

Music type	Attacks	Robustness
Pop	Cropping	1
	Amplification	1
	Echo	1
	Noise (AWGN)	1
	High bass	1

Table 6.8 Calculated values of SNR after applying different attacks for (pop music type)

Music type	Attacks	SNR
Pop	Cropping	27.0543
	Amplification	26.5142
	Echo	27.0543
	Noise (AWGN)	27.0543
	High bass	27.0543

Table 6.9 Calculated values of BER after applying different attacks for (pop music type)

Music type	Attacks	BER
Pop	Cropping	1
	Amplification	1
	Echo	1
	Noise (AWGN)	1
	High bass	1

Table 6.10 Calculated values of robustness after applying different attacks for (rock type)

Music type	Attacks	Robustness
Rock	Cropping	1
	Amplification	1
	Echo	1
	Noise (AWGN)	1
	High bass	1

Table 6.11 Calculated values of SNR after applying different attacks for (rock music type)

Music type	Attacks	SNR
Rock	Cropping	28.5751
	Amplification	26.0025
	Echo	28.5751
	Noise (AWGN)	28.5751
	High bass	28.5751

Table 6.12 Calculated values of BER after applying different attacks for (rock music type)

Music type	Attacks	BER
Rock	Cropping	1
	Amplification	1
	Echo	1
	Noise (AWGN)	1
	High bass	1

Table 6.13 Calculated values of robustness after applying different attacks for (opera music type)

Music type	Attacks	Robustness
Opera	Cropping	1
	Amplification	1
	Echo	1
	Noise (AWGN)	1
	High bass	1

Table 6.14 Calculated values of SNR after applying different attacks for (opera music type)

Music type	Attacks	SNR
Opera	Cropping	28.5751
	Amplification	26.0025
	Echo	28.5751
	Noise (AWGN)	28.5751
	High bass	28.5751

Table 6.15 Calculated values of BER after applying different attacks for (opera music type)

Music type	Attacks	BER
Opera	Cropping	1
	Amplification	1
	Echo	1
	Noise (AWGN)	1
	High bass	1

Table 6.16 Calculated values of robustness after applying different attacks for (Indian classical music type)

Music type	Attacks	Robustness
Indian classical	Cropping	1
	Amplification	1
	Echo	1
	Noise (AWGN)	1
	High bass	1

Table 6.17 Calculated values of SNR after applying different attacks for (Indian classical music type)

Music type	Attacks	SNR
Indian classical	Cropping	30.1242
	Amplification	30.1242
	Echo	30.1242
	Noise (AWGN)	30.1242
	High bass	30.1242

Table 6.18 Calculated values of BER after applying different attacks for (Indian classical music type)

Music type	Attacks	BER
Indian classical	Cropping	1
	Amplification	1
	Echo	1
	Noise (AWGN)	1
	High bass	1

Table 6.19 Comparison of BER values with previous work for cropping attack

Cropping attack		
Author	Method	BER
Wang and Zhao (2006)	DWT-DCT	0(front 1s only)
Wang et al. (2007)	SVR-DWT-DCT	0.0542
Zhao et al. (2014)	SVD-DWT	0.2
Proposed	DWT-SVD-DFT	0

Table 6.20 Comparison of BER values with previous work for echo attack

Echo attack		
Author	Method	BER
Al-Haj and Mohammad (2010)	DWT-SVD	1.5325
Al-Haj et al. (2011)	DWT	0.0884
Wang and Zhao (2006)	DWT-DCT	0.0857
Wang et al. (2007)	SVR-DWT-DCT	0.0857
Bhat et al. (2010)	DWT-SVD	0.0884
Proposed	DWT-SVD-DFT	0

Table 6.21 Comparison of robustness values with previous work for cropping attack

Cropping attack		
Author	Method	Robustness
Al-Haj et al. (2011)	DWT	1
Wang and Zhao (2006)	DWT-DCT	1
Wang et al. (2007)	SVR-DWT-DCT	0.9747
Bhat et al. (2010)	DWT-SVD	1
Proposed	DWT-SVD-DFT	1

Table 6.22 Comparison of robustness values with previous work for echo attack

Echo attack		
Author	Method	Robustness
Al-Haj et al. (2011)	DWT	0.966
Darabkh (2014)	DWT-SVD	0.9574
Wang and Zhao (2006)	DWT-DCT	0.9474
Wang et al. (2007)	SVR-DWT-DCT	0.9747
Bhat et al. (2010)	DWT-SVD	0.9458
Proposed	DWT-SVD-DFT	1

Table 6.23 Comparison of robustness values with previous work for AWGN attack

AWGN attack		
Author	Method	Robustness
Al-Haj et al. (2011)	DWT	0.968
Cox et al. (1997)	DCT based	0.78
Wang and Zhao (2006)	DWT-DCT based	0.0115
Wang et al. (2007)	SVR-DWT-DCT	0.9996
Xiang (2011)	DWT based	5.875
Wu et al. (2005)	DWT based	7.525
Proposed	DWT-SVD-DFT	1

Table 6.24 Comparison of robustness values with previous work for amplify attack

Amplify attack		
Author	Method	Robustness
Al-Haj et al. (2011)	DWT	0.915
Darabkh (2014)	DWT-SVD	0.9574
Wang and Zhao (2006)	DWT-DCT	0.356
Ozer et al. (2005)	SVD-STFT based	0.375
Proposed	DWT-SVD-DFT	1

Table 6.25 Comparison of SNR values with previous work

Author	Method	SNR
Wang and Zhao (2006)	Echo	21.47
Uludag and Arslan (2001)	DC-level shifting	21.24
Al-Haj et al. (2011)	DWT	28.5525
Cox et al. (1997)	Spread spectrum	28.59
Swanson et al. (1998)	Frequency masking	12.87
Nikmehr and Hashemy (2010)	Adaptive DWT SVD	24.37
Vivekananda et al. (2010)	DWT and DCT	28.6
Xiang (2011)	DWT based	23.98
Baranwal and Dutta (2011)	Spread spectrum and DWT	23.5
Proposed	DWT-SVD-DFT	30.12

6.5   GUI Implementation of the Digital Audio Watermarking

This is the main screen when the software launches. Anyone can set the parameter according to their requirement (Fig. 6.34).

After applying the desired parameters, the next screen shows the original image with the encrypted key and the recovered image with encrypted key images of the watermark image (Fig. 6.35).

This particular image depicts the original audio file and the watermarked audio file with an encrypted image embedded within it (Fig. 6.36).

Table 6.26 Effects of various attacks and recovered watermark image for KP.jpg image (summary)

Sr. no.	Music type	Attacks	Robustness	Image recovered
1	Instrumental	Cropping	1	Recovered Image
		Amplification	1	
		ECHO	1	
		Noise (AWGN)	1	
		HIGH BASS	1	
		Cropping	1	
2	Jazz	Cropping	1	Recovered Image
		Amplification	1	
		ECHO	1	
		Noise (AWGN)	1	
		HIGH BASS	1	
		Cropping	1	
3	Pop	Cropping	1	Recovered Image
		Amplification	1	
		ECHO	1	
		Noise (AWGN)	1	
		HIGH BASS	1	
		Cropping	1	
4	Rock music	Cropping	1	Recovered Image
		Amplification	1	
		ECHO	1	
		Noise (AWGN)	1	
		HIGH BASS	1	
		Cropping	1	
5	Opera	Cropping	1	Recovered Image
		Amplification	1	
		ECHO	1	
		Noise (AWGN)	1	
		HIGH BASS	1	
		Cropping	1	
6	Indian classical music	Cropping	1	Recovered Image
		Amplification	1	
		ECHO	1	
		Noise (AWGN)	1	
		HIGH BASS	1	
		Cropping	1	

Table 6.27 Effects of various attacks and recovered watermark image for Checkboard.jpg image (summary)

Sr. no.	Music type	Attacks	Robustness	Image recovered
1	Instrumental	Cropping	1	**Exctracted Image**
		Amplification	1	
		ECHO	1	
		Noise (AWGN)	1	
		HIGH BASS	1	
		Cropping	1	
2	Jazz	Cropping	1	**Exctracted Image**
		Amplification	1	
		ECHO	1	
		Noise (AWGN)	1	
		HIGH BASS	1	
		Cropping	1	
3	Pop	Cropping	1	**Exctracted Image**
		Amplification	1	
		ECHO	1	
		Noise (AWGN)	1	
		HIGH BASS	1	
		Cropping	1	

(continued)

Table 6.27 (continued)

Sr. no.	Music type	Attacks	Robustness	Image recovered
4	Rock music	Cropping	1	Exctracted Image
		Amplification	1	
		ECHO	1	
		Noise (AWGN)	1	
		HIGH BASS	1	
		Cropping	1	
5	Opera	Cropping	1	Exctracted Image
		Amplification	1	
		ECHO	1	
		Noise (AWGN)	1	
		HIGH BASS	1	
		Cropping	1	
6	Indian classical music	Cropping	1	Exctracted Image
		Amplification	1	
		ECHO	1	
		Noise (AWGN)	1	
		HIGH BASS	1	
		Cropping	1	

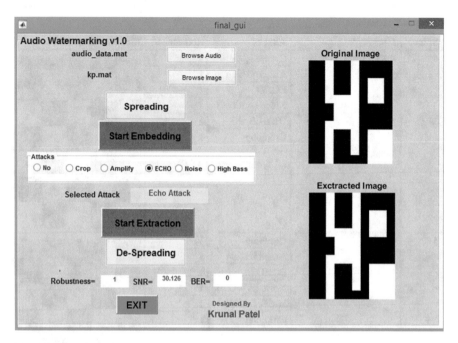

Fig. 6.34 MATLAB implementation—(GUI of Audio Watermarking Software)

References

Al-Haj A, Mohammad A (2010) Digital audio watermarking based on the discrete wavelets transform and singular value decomposition. Eur J Sci Res 39:6–21

Al-Haj A, Mohammad A, Bata L (2011) DWT-based audio watermarking. Int Arab J Informat Technol 8:326–333

Baranwal N, Dutta K (2011) Peak detection based spread spectrum audio watermarking using discrete wavelet transform. Int J Comput Appl 24(1):16–20

Bhat VK, Sengupta I, Das A (2010) An adaptive audio watermarking based on the singular value decomposition in the wavelet domain. Digital Signal Process 20:1547–1558

Cox IJ, Kilian J, Leighton T, Shamoon T (1997) Secure spread spectrum watermarking for multimedia. IEEE Trans Image Process 6:1673–1687

Darabkh KA (2014) Imperceptible and robust DWT-SVD-based digital audio watermarking algorithm. J Softw Eng Appl 7:859–871. https://doi.org/10.4236/jsea.2014.710077

Nikmehr H, Hashemy ST (2010) A new approach to audio watermarking using discrete wavelet & cosine transforms. Int Conf Commun Eng 183:1–10

Ozer H, Sankur B, Memon N (2005) An SVD-based audio watermarking technique. In: ACM workshop on multimedia and security. ACM, New York, pp 51–56

Swanson MD, Zhu B, Tewfik AH, Boney L (1998) Robust audio watermarking using perceptual masking. Signal Process 66:337–355. https://doi.org/10.1016/S0165-1684(98)00014-0

Uludag U, Arslan L (2001) Audio watermarking using DC-level shifting. Project Report. Bogazici University, Istanbul. http://www.busim.ee.boun.edu.tr/speechweb/

Vivekananda BK, Sengupta I, Das A (2010) An adaptive audio watermarking based on the singular value decomposition in the wavelet domain. In: Digital signal processing. Elsevier, Amsterdam, pp 1547–1558

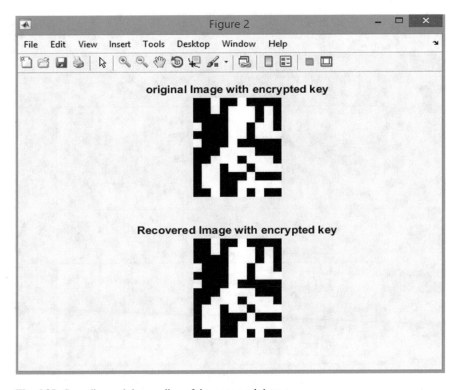

Fig. 6.35 Spreading and despreading of the watermark image

Wang X, Zhao H (2006) A novel synchronization invariant audio watermarking scheme based on DWT and DCT. IEEE Trans Signal Process 54(12):4835–4840

Wang X, Qi W, Niu P (2007) A new adaptive digital audio watermarking based on support vector regression. IEEE Trans Audio Speech Lang Process 15(8):2270–2277

Wu S, Huang J, Huang D, Shi Y (2005) Efficiently self-synchronized audio watermarking for assured audio data transmission. IEEE Trans Broadcast 51(1):69–76

Xiang S (2011) Audio watermarking robust against D/A and A/D conversions. EURASIP J Adv Signal Process 3:1–14

Zhao H, Wang F, Chend Z, Liu J (2014) A robust audio watermarking algorithm based on SVD-DWT. Elektronika Ir Elektrotechnika 20(1):75–80

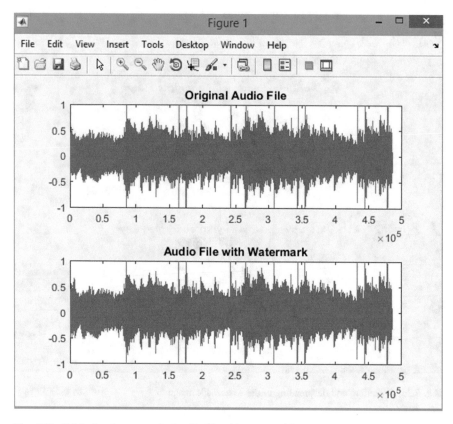

Fig. 6.36 Original and watermarked audio file with encrypted image

Chapter 7
Conclusion and Future Work

Abstract In this chapter, overall summary of the proposed work is described and also the scope of future work in this domain is discussed. At last in the bibliography section, all the papers which are referred for this work are listed and also the publication information for this research work is given.

Keywords Robustness · Security · Imperceptibility · DWT · SVD · DFT · DSSS

7.1 Conclusion

The expanding measure of advanced data that is accessible for duplicate downloads on computers and transmitting carefully through web stresses the requirement for copyright insurance. This issue has influenced the licensed innovation proprietors to look for an answer to stop copyright infringement. Digital watermarking gives a defense against such issues. As of late a few strategies for digital watermarking of electronic data have been created and presented by specialists. Extensive quantities of papers and electronic letters have been submitted in the writing depicting distinctive strategies and procedures in the watermarking field. In any case, this early research field is as yet open for broad work as technology progresses with a section of time.

In this book, new algorithm has been developed to embed watermark image in the audio files. The proposed algorithm is based on four-level DWT, SVD, and DFT transformations. The schemes introduced secret key embedding using DSSS scheme in conjunction with the watermarking method to increase the security and better copyright protection.

Performance evaluations were completed to demonstrate the value of the new algorithm. An extensive variety of scaling factors was analyzed to set up robustness and imperceptibility. Exhaustive tests were executed utilizing diverse assaults to assess the algorithm to affirm robustness. Distinctive insights were inferred, including the SNR, robustness, and BER to grill the watermarked audio quality and decide the comparability between the original audio file and the reproduced audio file. The closeness between the original watermark and the recovered watermark was additionally completely decided to utilize the BER measurement. The trial results affirmed that the new algorithm showed high estimations of robustness, SNR, and BER, which infer a high perceptual transparency of the watermarked image.

The abstract perception was additionally considered to approve of the new algorithm and demonstrated positive proof. The BER value estimates how intently the embedded watermark coordinates the extracted watermark. A BER equivalent to 0 implies perfect discovery without any error. From the test results, the BER values were quite often high inferring that the algorithms are robust against common watermarking attacks. The execution of the new algorithms was additionally checked by examinations with different techniques in the writing to affirm prevalence. Such assessments were performed utilizing diverse audios utilized in conjunction with watermarks of various measurements.

The experimental results suggest strongly that the new algorithms are a significant improvement over standard schemes. The simulation results in term of robustness show that the new algorithms have a superior performance in many instances. In most of the cases, results show that 100% watermark detection is guaranteed whilst maintaining a high quality of SNR.

This experiment has been conducted in the laboratory under a controlled environment using MATLAB 2015b. It is likewise vital to take note that there is nothing to propose that a similar guideline may not work in the real ongoing condition and give comparable outcomes. Be that as it may, this should be contemplated in real field preliminary.

The above examinations demonstrate that, by and large, the execution of the proposed algorithms is superior to other watermarking techniques. In summary, great certainty can be put on the results of this examination, which is both unique and noteworthy from the scientific or programming designing point of view.

7.2 Future Work

There are a few zones of this work that would profit by advanced development and examination keeping in mind the end goal to enhance the performance. Despite the fact that the plans exhibited are very straightforward and powerful against numerous audio-processing attacks.

The audio watermarking is moderately new and has a wide degree of investigating. This proposition is restricted to binary image embedding and can be kept on turning grayscale images. The procedure can be executed on live signals as opposed to a fixed signal as considered in this proposition. Additionally, research can be carried on inserting watermark in video, that is, movies or surveillance systems. Applying watermarking procedure on surveillance will diminish the security issues by monitoring voice communication. One other application that can be focused on is the watermarking of the live recording, for example, a man recording his video and capturing images.

Such multimedia applications could be implemented on hardware platforms. Real-time hardware implementation could offer advantages to images acquired using mobile phone cameras. It would enable processors on cell phones to be sufficiently quick with the goal that they can watermark pictures, audios, and video progressively.

Index

© The Author(s), under exclusive license to Springer Nature Switzerland AG 2021
K. N. Patel, *Robust and Secured Digital Audio Watermarking*, SpringerBriefs in
Speech Technology, https://doi.org/10.1007/978-3-030-53911-5